D1282596

PHILOSOPHERS OF THE EARTH

Philosophers of the Earth

Conversations with Ecologists

by

Anne Chisholm

We used to be studying guillemots: now we are trying to save the human species

—DR NORMAN MOORE

All thinking worthy of the name must now be ecological

—LEWIS MUMFORD

SIDGWICK & JACKSON

LONDON

ISBN 0 283 97851 1

Printed in Great Britain
at the St Ann's Press, Park Road, Altrincham
Cheshire, WA14 5QQ
for Sidgwick and Jackson Limited
1 Tavistock Chambers, Bloomsbury Way
London WC1A 2SG

To my parents

CONTENTS

ILLUSTRATIONS

ACKNOWLEDGEMENTS

To write this book, I have had to make repeated demands on the good will and patience of a number of busy people. I should like to thank first of all those whose conversation and activities make up the substance of the book, especially all those named in the chapter headings. All of them found time to talk to me at length, and then to read and correct their own contributions. Any mistakes which remain are entirely mine. I should like to thank in particular Sir Frank Fraser Darling, whose writing and conversation led me to become interested in ecology, and who encouraged and helped me at the beginning of the project.

As well as those quoted in the text, many other people gave me advice and information which was invaluable. I should especially like to thank Mr Max Nicholson, the first head of the Nature Conservancy, and Dr Duncan Poore, the present director. I received much help from everyone I met at the Monks Wood Experimental Research Station. I should also like to thank Mr David Brower of Friends of the Earth, U.S.A., Dr Gerardo Budowski, director of the International Union for the Conservation of Nature, Dr Stanley Cain, director of the Institute for Environmental Quality at the University of Michigan, Mr Robert C. Cook, formerly director of the Population Reference Bureau, Washington D.C., Dr Marshall I. Goldman of the Russian Research Centre, Harvard, Mr Gordon Harrison of the Ford Foundation, New York, Mr Gordon Howe, director of the Conservation Foundation, Washington D.C., Mr Russell Train, Chairman of the United States Government's Council for Environmental Quality, and Dr George Woodwell of the Brookhaven National Laboratory. I also want to thank Jeremy Swift for his advice and guidance, and Michael Davie for his encouragement and support.

FOREWORD

LIKE a lot of other people, I first heard the word ecology in the autumn of 1969. This was shortly before that year's series of Reith Lectures on the B.B.C., which were to be given, it was announced, by Dr (now Sir) Frank Fraser Darling, an ecologist. As a journalist, I was asked to write about the lecturer. I looked up ecology in the dictionary, and found that it was the science of the inter-relationships between living organisms and their environment. Not feeling much wiser, I then read the drafts of his lectures.

Suddenly, I discovered a new subject. Ecology, it appeared, was the science which could interpret the fragments of evidence that told us something was wrong with the world – dead birds, oil in the sea, poisoned crops, the population explosion. Ecology, it seemed to me at that stage, was an especially enjoyable abstract idea. What it meant was – everything links up. What interested me, first of all, was not the actual content of ecology but the ecological message. Here, it seemed, was a new morality and a strategy for human survival rolled into one.

I soon realized that this revelation was not, to put it mildly, mine alone. During 1969 and 1970, ecology became fashionable. In Britain, both the major political parties appointed Ministers of the Environment. (All the same, Reginald Maudling, the Conservative Shadow Minister, was heard to enquire, while the Reith Lectures were on, if someone could please tell him, what *was* ecology.) Not long afterwards, the Labour Government announced a Royal Commission on Environmental Pollution. In America, as usual several steps ahead, President Nixon's State of the Union message featured a plea for a new attitude to the environment; an Environmental Protection Act was passed; a Council for Environmental Quality was formed. During the autumn of 1969, ecology caught on like a new religion among the young on college campuses across the country. In March, 1970, came the celebration of Earth Day. In London and New York, health food stores, hitherto usually considered a cranky fad, became big business. In Europe, 1970 was European Conservation Year. Newspapers, magazines, and television started making the environment big news. The B.B.C. ran a popular television series called *Doomwatch* in which a handsome Nobel-prizewinner civil servant saved the country once a week from unforeseen ecological catastrophes. Inevitably, advertising followed. Oil and chemical companies began to tell us how concerned they were about pollution and the quality of life. Suburban housewives began to worry about using too much detergent and whether they should re-cycle their garbage.

Confusion set in, as the words ecology, environment, conservation, biosphere, became part of everyday language for the first time. It was accompanied by a generalized sense of alarm as anxious scientists, whom the media labelled Prophets of Doom, gave us assorted and sometimes contradictory warnings. The earth was heating up and the ice caps would melt and flood the world. The earth was cooling down and a new Ice Age would be upon us. We all have D.D.T. in our fat: was this harming us or not? The population of the world would inexorably double in thirty years: would our resources last? The people who were supposed to have the answers were the ecologists.

They could, it seemed, offer a general philosophy of life that explained man's dependence upon, and responsibility towards, nature. They were trained in a specialized field of study that could instruct us in how the whole natural system worked, tell us what we were doing wrong, and advise us how to put it right. If you believe that everything depends upon everything else, then you believe that damage to the smallest part of the great machine may have serious consequences. In an increasingly directionless, secular world, the intricate patterns traced by the ecologists offered a potential source of comfort, a sense of the unity and beauty of life. The neatness of these patterns was pleasurable in itself, and curiously reassuring. Charles Darwin was responsible for a classic ecological joke that connected the passion of English spinsters for cats with the greatness of the Royal Navy. He started by looking closely at red clover and the bumble-bees which pollinate the flowers. These bees, it appeared, were the only insects which could pollinate red clover successfully, because of their specially long tongues. So the clover depends on the bumble bee. It had been observed that there were more bees near towns and villages than deep in the countryside, because field mice, which eat honeycombs and larvae, were scarcer near civilization. Why were the field mice scarce? Because towns and villages have large cat populations, which keep their numbers down. A German scientist took the theory several steps further. Cats contributed to the success of red clover in Britain; British cattle fed on red clover; the British navy fed on beef from those cattle; hence the cats were responsible for Britain's success as a naval power. Thomas Huxley, Darwin's supporter, suggested that since it was the old maids of Britain who chiefly supported the cat population, it was, in fact, the British spinster who kept the Royal Navy great.

Moreover, the ecologists offered good reasons for taking seriously incidents which hitherto had caused the general public little concern beyond a vague distaste, a sensation of aesthetic displeasure. The ugly litter of plastic bottles on the beach became something more when you heard that the frozen edges of the Arctic or Antarctic, were becoming a global rubbish dump. What would happen to the world's garbage if it never decayed? If thousands of dead seabirds had been poisoned by chemicals in rivers and oceans, what about the fish we ate and the water we drank? If pelicans in California and herons in East Anglia broke their own eggs and ate them, and pesticide residues were suspected of having upset some instinctive avian mechanism, what were the implications for human beings?

So nature conservation, too, began to take on a new importance. Hitherto, the saving of stretches of unspoiled country, or the rescue of threatened eagles or elephants, had probably seemed a good idea to most people, if they gave it any thought at all, but conservation was hardly a top priority. Now, the ecologists were talking as if conservation was central to man's own well-being, possibly to his survival. So the question arose: was conservation simply applied ecology? And how exactly did saving a species of animal or the creation of a national park help man to solve his environmental ills? Again, the ecologists were supposed to know the answer.

I started by assuming that I could tell the story of ecology and of the conservation movement simultaneously. But it did not take long to find out that the conservation story needs a book, or several books, to itself. Many of the people who have worked hardest, and achieved most, in getting National Parks or Nature Reserves established are not ecologists or even biologists, though they are usually good amateur naturalists. I decided therefore to stick to ecologists, or important ecological thinkers, dealing with those aspects of the conservation movement that emerged naturally as I asked each person about his or her ideas and career.

Along the way, I encountered a number of other fundamental questions. Ecology may be a new subject for most of us – and in the sweeping way it has been applied to the human environment in the last two or three years, it is indeed new – but I soon found out that it has a history. I realized, too, that not all ecologists welcome the sudden popularity of their subject. For them, ecology is a specialized form of biology, based on minute studies of plant and animal communities and intricate calculations about their interrelationships, in which man and his clumsy activities have no place. There is a tradition of argument among ecologists as to whether a science of 'human ecology' exists in its own right, or whether the term is just another name for sociology or demography. But two points remained clear: man's actions are impinging more and more on the natural world; and ecologists, whether they like it or not, have the knowledge and the insight to help him understand what he is doing. I also discovered that ecologists have occasionally been criticized for failing to rise to their opportunities, for being content to stay in the background, and for not realizing their responsibility towards the rest of us. Max Nicholson, the ornithologist and leading British and international conservationist who was largely responsible for founding the Nature Conservancy in Britain after the war, wrote in his book *The Environmental Revolution* (Hodder, 1969): 'The impact of ecology upon the attitudes and practices of mankind has hitherto been disappointingly limited, and far below the obvious potential of the contribution which it could and should make. . . . Far from breeding a race of benevolent superminds modern ecology has either been content or compelled to accept a lowly role on the sidelines of biology . . . Ecologists must face the fact that they have largely failed to justify and turn to advantage the immense goodwill which ecology in the abstract commands among thinking people'. Were such criticisms justified? Now that ecology was in the news, were the ecologists seizing their chance?

Above all, I have tried to discover what ecological thinking is, and what ecologists actually do. I have traced the development of their careers and their ideas in some detail, because I found that their life-stories seemed to mirror the expansion of the subject. How does it happen that a young ecologist starts by studying a herd of red deer, and forty years later is offering a total philosophy for modern man?

The first section of the book deals with the background to our new environmental anxieties, and describes some of the older men who have been working out their ideas for forty years or more. It includes three men who are not ecologists, but whose ideas are strongly ecological, and who in my view have made vital contributions to the development of ecology as an attitude of mind. The second section deals with scientists who are putting these ideas into practice in their day to day work. The third section is devoted to some of the activists who have been using their time and energy to arouse and involve the public. And the fourth section looks at the ideas of four men whose work seems to me to point the way ahead. Naturally these categories sometimes overlap.

My aim, in short, has been to allow a selection of the people whose ideas and activities have formed and are forming ecology to explain, in their own way, the range and substance of an extraordinarily varied, complex, and crucial subject.

I

Lewis Mumford

INVENTING THE ENVIRONMENT

OF ALL the wise men whose thinking and writing over the years has helped to prepare the ground for the environmental revolution, Lewis Mumford, the American philosopher and writer, must be pre-eminent. His name is most closely associated with contemporary architecture, town-planning, and the history and shape of cities. Since the early thirties he has been recognized as the hammer of indiscriminate technology. Now in his mid-seventies, his latest book (he has written some fifteen during the last fifty years) *The Pentagon of Power,* is a passionate indictment of what he has called the Megamachine, the whole technological complex that modern man has constructed but not yet learned to control. He has studied and written little about ecology as such, but in the broadest sense man's environment has been his main concern from the start.

Mumford describes himself as a New Yorker born and bred, and his home is in the country north of New York City, but for several years he has spent the autumn terms in Cambridge, Massachusetts, as a visiting scholar at Leverett House, Harvard. He lives at the top of a modern residential college building near the river where he and his wife have a penthouse apartment, with a good view over the university, modern furniture, and an atmosphere of friendly restraint. He is a tallish man, very upright, looking much less than his seventy-six years. He has a fine, imperious head, a ruddy complexion, a neat grey moustache, keen, bright eyes, and a courteous, kindly manner.

Surprisingly, in view of his eminence and the relevance of his life's work to the new concern with the environment, few people in Cambridge seemed aware of his presence and students did not queue up to tap his wisdom. Harvard seemed an utterly appropriate place for him to be, for he has the style of an upper-class East Coast academic, but to my astonishment I had discovered that Mumford has no formal degrees of any kind, apart from honorary ones, though he has studied and taught at several of America's great universities, including Stanford, California, and the University of Pennsylvania. He has visited England frequently and loves and admires the country.

He was born in Flushing, Long Island, in 1895, where his father was a lawyer. As a boy, Mumford went to a technical high school and intended to become an electrical engineer. The first article he ever wrote was about radio sets, and during the First World War he was a radio electrician, second class, in the American Navy. It is a surprising background for someone who was to become one of the true pundits of his age, a dominant influence on a whole generation of post-war architects and town planners, and one of the great generalists. Nowadays, Mumford is acknowledged as a prophet in his own country; but while greatly honoured, his position is curious. He is recognized as an outstanding writer and thinker, but he is remote from the arena of fashionable intellectual debate.

Thus, the suggestion that ecology has only recently become a popular science cut no ice with Mumford. 'Ecology has always been part of my thinking', he said firmly. 'My interest in it dates back to my early interest in Patrick Geddes.' (Sir Patrick Geddes, the Scottish biologist and planner, was a personal friend as well as a mentor of Mumford, who named his only son after him. Geddes Mumford was killed in Italy in the Second World War.) 'I have an undue reputation for having promoted ecology; I have never promoted ecology as such. It has just been part of my life; it hasn't been an independent subject. I think about the entire complex, the whole environment, not in terms of any one fragment of it. That is ecological thinking, but I'm not thinking about ecology when I'm doing that, I'm thinking about something else.'

Anxious to keep my two working definitions of ecology straight, I tried them out on Mumford: 'Is not ecology both a specific scientific study *and* an overall awareness of "the entire complex"?'

'It really begins with a combination of two things, science, and a sense of responsibility to the world of life. That begins with Darwin, and I have a passage in my new book about Darwin, not as the discoverer of evolution – a lot of people have "discovered" evolution – but as the supreme ecologist.' When I referred to the passage on

Darwin in *The Pentagon of Power,* I found he rated ecology very highly indeed. 'A growing appreciation of all that distinguishes the world of organisms from the world of machines gave rise, in the nineteenth century, to a fresh vision of the entire cosmic process. This vision was profoundly different from the one offered by those who left out of their world picture the essential qualitative attribute of life; its expectancy, its inner impetus, its insurgency, its creativity, its ability under exceptional circumstances to transcend either physical or organic limitations. The name given to this new vision of life was bestowed belatedly, only when it began to be systematically pursued: it is ecology.'

The primary theme of *The Pentagon of Power,* which like all Mumford's writings is packed with learned and enlightening disquisitions on literature, anthropology, science, and the history of man's social and intellectual attitudes, is that since the seventeenth century man's urge to dominate nature and his fellow men has led him astray. Power over nature was acquired at a cost. Because both the method and the ideology depended on breaking down natural phenomena into manageable parts, theorizing, experimenting, then moving on to the next problem, men lost the sense of life as a great web, which is what ecology teaches. Mumford's thesis is that gradually man came to despise the world of nature, to feel that his mind had outflanked it and that his machines could take over its essential functions. The study of the ecology of natural systems, he indicates, is the crucial counterbalance to this arrogant view. But such a study is not simple. 'The ecological complexities of existence overpower the human mind, even though some of that richness is an integral part of man's own nature. It is only by isolating some little part of that existence for a short time that it can be momentarily grasped. We learn only from samples. By separating primary from secondary qualities, by making mathematical description the test of truth, by utilizing only a part of the human self to explore only a part of its environment, the new science successfully turned the most significant attribute of life into purely secondary phenomena, ticketed for replacement by the machine. Thus living organisms in their most typical functions and purposes became superfluous.'

If the ecological approach is so crucial, I asked, how do we achieve it?

'First of all I think it requires a sense of the way life functions,' said Mumford. 'No part of life can be considered apart from any other; there is no such thing as an organism without an environment, just as there's no such thing on earth as an environment without an organism. Even the most barren environment has organisms; you can always find them if you look hard enough. This interaction goes on all the time; this is what life is. People who have been trained purely in the physical sciences often lack this sense; and as they become specialized, technologically, they separate a very small part of the reality from the whole. They deal with that competently in its own right, but they lose sight of the whole. So really we need a reorientation of our education if we are to be thinking ecologically.'

But is it not essential, I asked, for the ecological thinker to acquire a firm base in one particular discipline? I was speaking in general terms, but Mumford related my question to his own intellectual development.

'Well, I've done that in more than one discipline as a matter of fact and perhaps it has been a help. I have a reputation as a specialist in Utopias, in American culture, both in literature and architecture. I was one of the pioneers in the field of American studies, which simply weren't taught in many colleges until the thirties. And I have a reputation as a historian of technology and a student of cities. In each case I have a specialist's reputation, because I have studied these fields sufficiently to know a little more than most people. So I never recommend anybody to study things in general. You must know at least one thing well, and have access to the same kind of knowledge in other fields. But then combining them together is a habit of life.'

Mumford's early and abiding interest in town planning and urban life was clearly fundamental to the way his thinking developed. 'There was', he explained 'a group of us in New York, back in the twenties. Clarence Stein, the architect and planner, and Henry Wright – the people who planned Radburne, the first American attempt to create a garden city; and Benton Mackaye, a forester and conservationist, who conceived and helped establish the Appalachian Trail; and Stuart Chase, who was an economist; all of us had a particular interest in cities and a general interest in improving the whole environment. We laid the foundation for the whole regionalist movement, in housing and community planning, which arose here in the thirties, and afterwards, although it didn't follow our line, in fact it departed from it very seriously. Nevertheless, it had its beginning there. When Roosevelt was in office he founded State Planning Boards, and they were, in effect, regional planning centres. This system wasn't carried through, but it was the result of the preliminary work we had done. There was an enormous amount of political opposition to it – jealousy between the states who hadn't yet learned to co-operate with each other. Also there was considerable distrust of planning in any form. Planning was something the government was going to do to you. The way people in democracies think of the government as something different from themselves is a real handicap. And, of course, sometimes the government confirms their opinion, unfortunately.'

Mumford combined his interest in practical planning with a study of ideal communities. His first book, in 1922, was a study of Utopias, and his later writing often returns to this theme. Consistency is a Mumford attribute. Nearly all the themes in his most recent writing can be found, germinating, forty or fifty years ago, I unearthed a yellow clipping from the London *Times* of 1927 in which a reviewer of an early Mumford book, *The Golden Day* (Cape, 1929), summarized its main subjects thus: 'Economics, Rousseau, the perfectibility of mankind, the pioneer, the settlers in America, the taming of nature, the industrial age, hopeful writers, hopeless writers'. He suggested then that a new philosophy might come about which, he wrote, 'shall be oriented as completely towards Life as the dominant thought since Descartes has been directed towards the Machine'. *The Pentagon of Power,* which came out in 1970, explores and illustrates this very idea.

As a student of cities who had spent much of his life in and around New York, Boston, Philadelphia, and London, Professor Mumford has experienced at first hand the coming of the Urban Nightmare. Recently he has started referring grimly to 'Kakotopia' – the opposite of Utopia – which he has defined as 'a misplanned and ugly urbanoid place'. I suggested that he must have found it painful to watch the horrors accumulate over the last fifty years. Professor Mumford sat up even straighter and spoke with feeling. 'The pain gets worse and worse as time goes on. For instance, I saw what was coming before the National Highway Act was even passed. I saw it out in Pittsburgh in the fifties. They began thrusting superhighways right into the heart of the city, and I knew this would destroy Pittsburgh and I said so. What infuriates me is that the city planners, who ought to have been on their toes, accepted this as the natural form of population dispersal. They didn't attempt to fight it. One is furious over the stupidity of people who are professionally trained and still don't see the immediate issue.'

Perhaps, now that we are all more aware of what is going on, the professionals won't be able to make the same old mistakes?

'Well, they're still making them. The thing that's correcting them is the revolt on the part of the people affected. I get letters from all over the country, saying, "Won't you help us fight this highway that's eating into the heart of our town, or going to destroy a beautiful landscape". My answer to them is, you've got to fight it yourselves, my authority doesn't count for anything. If you do it, then your elected representatives and even the corporate interests involved will have to pay some attention to you. The revolt is beginning to have some effect. They've actually stopped the butchering of one Boston neighbourhood by a highway. An enormous amount of federal money has already been spent on developing a system of highways which is wrecking both the landscape and the cities. It is only local opposition, often by a persistent minority, that is preventing this from happening.'

Here Mumford became noticeably excited over the political implications of public action and protest, and I remembered that despite his recent appearance of an Olympian aloofness from the political arena, he was a leading critic of the atomic and hydrogen bombs, and in 1965 startled an outraged part of the intellectual establishment when, in his speech of retirement from the august position of President of the American Academy of Arts and Sciences, he launched a blistering attack on U.S. foreign policy in the Dominican Republic and Vietnam. He is also credited with being the first public figure to write an open letter to President Johnson to protest against the bombing of North Vietnam. In fact, he is a senior protester.

'One of the things the student protest movement has shown is that if things get bad enough they can be stopped – if not by rational means then by irrational means.' Mumford bristled fiercely, like an infuriated elderly sealion. 'This might persuade the authorities to listen to people when they talk to them. The great thing is – I discovered this long ago in my own life – politics is a fascinating field once you get into it, and then it can become almost a full-time occupation. We now have a population which has almost as much leisure as the upper class population of Athens used to have; therefore we have people who have time to attend to politics. Until people were released from overwork they didn't have that time. This may change the whole nature of our political life, as soon as people can see that there is an opening for them.'

Despite the rumblings and warnings that Mumford and a few likeminded people were making over the years, it was not until the public began to be directly affected by a deteriorating environment that the message began to filter through. Mumford acknowledges this point with no apparent bitterness. 'Now that people get daily reports on radio and television about the amount of pollution that's in the air, they are beginning to think about what's happening to them. The situation reached a point where it could no longer be ignored. When you are caught in stalled traffic on a San Francisco bridge and realize that the gasoline fumes are giving you a headache, and making you feel ill, that causes you to think. It's the experience of pollution that's done it. You've only got to taste the water to know that the water supply is contaminated. There is a threshold beyond which everyone knows that pollution must be fought.'

In *The Pentagon of Power* Mumford places great emphasis on the importance of environmental deterioration in bringing home to mankind the mess we are in. 'Admittedly, the disasters of war, though no longer locally limited, had through the ages grown too familiar to bring about a sufficient reaction. During the last decade, fortunately, there has been a sudden, quite unpredictable awakening to prospects of a total catastrophe. The unrestricted increase of population, the over-exploitation of megatechnical inventions, the inordinate wastages of compulsory consumption, and the consequent deterioration of the environment through wholesale pollution, poisoning, bulldozing, to say nothing of the more irremediable waste products of atomic energy, have at last begun to create the reaction needed to overcome them. This awakening has become planet-wide. The experience of congestion, environmental degradation, and human demoralization now fall within the compass of everyone's daily experience. The extent of the approaching catastrophe and its dire inevitability, unless counter-measures are rapidly taken, has done far more than the vivid prospects of sudden nuclear extinction to bring on a sufficient psychological response. In this respect, the swifter the degradation, the more likely effective measures against it will be sought.'

But practical measures alone are not enough, Mumford strongly feels. He has always emphasized the need for human, spiritual renewal as well as action, and this message has reached a crescendo in his latest writing, much of which has the tone of an impassioned sermon. 'People are relying on technological devices to alter the ecological situation and I think they are mistaken', he said. 'They need to change their attitude towards themselves. Most of the really important, really decisive changes will have to be human changes, not technological changes. We have to change our habits of life, our expectations. We've been trained to accept waste as a necessary part of economic prosperity. All our advertising is an attempt to make us consume more than we should. This is the kind of change that has to be made. It has nothing to do with technology. These are fundamentally moral changes, changes in human habits. People are beginning to feel the need for a change, the young especially. One of the most interesting things is that there is a considerable body of young people who are extremely conscious of this requirement.'

But Mumford is not wholly uncritical of the young. 'They are also capable of going in exactly the opposite direction. The famous Woodstock Festival was a devastating attack on the environment. Over a hundred thousand young people all converging on the same place – they produced more pollution than even New York could show in the same space of time! Think what all those motor cars did to the air, and think of the erosion of the soil.' But he sees other, more hopeful signs. 'The young in Cambridge, for example, have gone back to bicycle riding; some of them even ask their friends not to visit them in the country unless they go by public transportation or on a bicycle. They don't want them if they come by motor car.'

Was this because they were anti-machine?

'Well, it's anti-pollution, really. If the machine doesn't pollute, like the bicycle or the electric car, it is perfectly acceptable. It was pure technological backwardness that we abandoned the electric car. We abandoned it as late as 1915, when Ford got his cheap motor car going strong. Up till that time the delivery of goods from department stores in New York was made by electric cars. I remember them. And they were efficient; they went faster than any gasoline-powered car actually goes in the city today because there weren't so many vehicles. The motor car tempted us to limit all our transportation to one method, and that's nonsense. We've undermined public services that once were very well handled. In Britain, too, try to get somewhere in Wales without a motor car! You used to have a very good train service, but it was dismantled because it wasn't profitable. You might as well dismantle the postal service. It may not be profitable to deliver letters for a few cents to someone who lives ten miles from the nearest post office, but it's necessary.'

One can trace, in Mumford's work, an increasing scepticism about wholesale mechanization and automation which at times recently has approached outright hostility. Not surprisingly, he was more optimistic in the twenties and thirties than he is now about man's capacity to restrain his own inventions. He can be bitterly scathing in his references to these monsters and their acolytes: 'If the first stage in mechanization 5,000 years ago in the Pyramid Age of Egypt and Mesopotamia was to reduce the worker to a docile and obedient drudge, the final stage that automation promises today is a self-sufficient mechanical electronic complex that has no need even for human nonentities . . . Humanly speaking, the proper name for automation is self-inflicted impotence. . . . If we intend to provide for the survival of plants and men, we had better ask the bat-eyed priests of technology what on earth they think they are doing.' But statements like these are always set against passages in which he acknowledges the benefits of technological knowledge, and he insisted, during our conversation, that ecology need not be anti-technology or anti-industrialization as such.

'What we are demanding', he said, 'is a different kind of technology, one that is in a better relationship to life, not one that is concerned merely with increasing the Gross National Product or providing enormous profits for those in control of the system. This isn't being anti-technology; it's being in favour of another kind of technology that can continue in existence indefinitely. It's only by putting back into the system what you have taken out of it that you can have a system that is in equilibrium. Our lopsided technology is now in the process of breaking down, and must eventually undermine its own existence.'

Mumford's strong historical sense has led him to attempt a reinstatement of the value of pre-industrial technology: 'Western man's pride in his many real achievements in mechanization made him too prone to overlook all that he owed to earlier or more primitive cultures.' Time and again in his books he extols the inventiveness of neolithic and medieval man, and the harmony that prevailed between those early technics and the environment. Quoted out of context, such passages could give the impression of unrealistic romanticism, but this is not so when his ideas are taken as a whole. Many times, he emphasizes that machines alone cannot be evil: it is man who is at fault. His great fear is that in surrendering unconditionally to the automated power system, modern man has forfeited some of the inner resources necessary to keep him alive: 'I have no quarrel with either science or technics so long as they remain subordinate to organic functions and human purposes', he said. This fear, and his passionate repudiation of the whole structure of destruction, war, dictatorship, exploitative economics and political repression, has led him to say some hard things about his country. The ultimate Megamachine, he says, was the Nazi state, but during and after the Second World War the enemies of the Nazis became contaminated by the very thing they were fighting to destroy: 'In the very act of dying, the Nazis transmitted the germs of their disease to their American opponents – not only the methods of compulsive organization and physical destruction but the moral corruption that made it feasible to employ these methods without stirring effective opposition.'

Encountering statements like this, Mumford's isolated position in America today seems less surprising. These feelings put him well outside the establishment pale; but by temperament he could never be a Spock, a Chomsky, or a Robert Lowell, marching the streets with the young protesters. Besides, Mumford has always been ahead of the times, which is no way to win instant popularity, and his very broadness of mind and concern for style set him apart from day to day polemics.

Although he has never worked directly with nature or animals,

Mumford is profoundly sensitive to natural beauty; he is a keen gardener, and used to go hunting with his son. Did he, I wondered need regular and direct experience of nature? Is the wilderness important to him personally?

'I think it is. For a minority of us it is a very important experience and as such should be preserved even if only a minority enjoys it. One of the things that ecology teaches is the need for variety; the uniform, homogenized environment is the most deadly of all. The part of the country I live in is important to me. We've had a country home there for forty years or more. I live in Dutchess County. It is pastoral country, and still produces milk that goes down to New York. It's a combination of meadowland, parkland, and woods. I find it very nourishing. One of the conditions of my life is that I have access to this landscape for at least part of the year. I haven't ever gone off into the wilderness for any length of time. I haven't felt the inner need for that, though some of my friends like MacKaye have. I've only walked a few miles on the Appalachian Trail, though I was in on its very inception. That kind of solitude I don't really need. But I do need solitude every day, and some contact with nature in a more or less primitive form, if only in the form of pasture.'

Among professional ecologists and environmentalists, Mumford has long been recognized as a kind of senior guru. He has several times taken part in their symposia and conferences as the man they think best equipped to sum up, and have the final word. In this capacity, he played a leading part in the Princeton Conference of 1955 on 'Man's Role in Changing the Face of the Earth', a gathering which resulted in a massive tome that has become required reading for environmental students. 'It was I who proposed the section on the future in that conference', Mumford told me with some pride. 'They had actually put together in their minds the papers for the conference, but they hadn't dealt with the future at all, and I said, "I don't belong here, because the future is my province". And so we arrived at a whole section on the future.' I mentioned that Sir Frank Fraser Darling had told me the conference was the most intensive intellectual experience he'd ever had. 'Mine too', Mumford said with enthusiasm.

Mumford greatly likes and respects ecologists like Sir Frank, who combine scientific knowledge and personal sensitivity. 'That characterizes most of the good people who've gone into ecology, as it characterized Darwin himself', he said reflectively. 'When he wanted to perform an experiment on a baby, he didn't take the baby into a laboratory, he put the baby in a girl's arms, and then performed the experiment. That's the kind of mind which is involved. Darwin knew the natural environment of a baby to be in a woman's arms, not on a laboratory table.'

Before I left, I asked Mumford whether he was still hopeful that the profound change in attitudes he deems necessary if man is to save himself is possible. He spoke with weary simplicity. 'Why, if one gave up hope one might just as well give up living. I'm always hopeful because there are many possibilities still open to us. But I'm a pessimist about probability, because there are many forces which will resist any sufficient change in our habits to the bitter end. We don't know yet which is going to win out. I don't make any bets on that.'

2

Rene Dubos

DIAGNOSIS OF A GENERAL SICKNESS

ANOTHER MAN who, although not a professional ecologist, has dramatically affected ecological thinking is Professor Rene Dubos, Emeritus Professor of Environmental Biomedicine at the Rockefeller University in New York city. He was born in France, but emigrated to America as a young man, and made his name as a bacteriologist in medical research in the thirties. He has a great reputation as a scientist, stemming from pioneering work on antibiotics. He also writes books for the general reader, one of which won a Pulitzer Prize, and regularly travels the lecture-circuit. He started writing and talking about ecology and the environment over twenty years ago.

Because he began his career by concentrating on human health, Dubos looks at the environment in terms of man's well-being. To him, the solemn debate among ecologists as to whether there can be a science of human ecology seems absurd. To think about the environment without assuming that man is the most crucial part of it is to him positively unnatural. His field of study has been man in his surroundings, and the effect of each upon the other. Naturally, all disciplines and philosophies are relevant, and Dubos deals in aesthetics, politics, and sociology, as well as medicine and environmental health. He likes nothing better than to generalize and build bridges between disciplines. His dominant theme is that if we want to improve our physical and spiritual well-being, we must first understand and then control our impact on our surroundings.

Dubos is in his early seventies; large, substantial, and slightly hunched about the shoulders; but it is still easy to imagine him as he has described himself as a young man: 'I am six feet tall and have blue-green eyes and Viking-like flaxen hair'. Now, he has a domed bald head with a fuzz of grey hair towards the back, and a strong face like a Breughel peasant. Until he speaks, he could pass for native American but his accent is as French as Maurice Chevalier's and his manner as charming, though less stagey.

I saw Dubos in his office at Rockefeller University, which, entirely unknown to many New Yorkers, is tucked away round a surprising, parklike campus on the East River in mid-town Manhattan. The professor works on the fourth floor of a modern block of science buildings, with white-coated laboratory workers in the corridors. His office is unremarkable; a big window looks uptown across grass and paved pathways and the walls are solid with books.

I asked Dubos first what he thought about the public's sudden interest in environmental matters.

'Do you realize what a division, an upheaval this issue of the environment has brought about in American scientific life?' he at once exclaimed. 'If I speak at a meeting, Lamont Cole and Odum (two distinguished professional ecologists), whom I'm sure respect me, will say, "But you talk about *man,* you do not talk about ecological systems". What they are interested in is the ecological system achieving an equilibrium of its own through natural forces. I stand for the extreme among the people who say there is no "natural" ecology; man has changed everything in nature.'

To Dubos, it is self-evident that to control our environment we must first understand man's nature and history, and he frequently uses England as an example of what he means.

'There is not a blade of grass in England which is not man-made. England was covered by forest until neolithic times, and so was the whole of continental Europe. What we call England, or France, or Germany, was created by neolithic farmers followed by medieval farmers and monks, who destroyed the forest and drained the swamps. All this land is man-made. If you take this view – and you cannot avoid it, those are the facts – you realize that when professional ecologists try to discuss ecological systems without recognizing man as the main component of the system they are not talking about reality.

In our new-found concern for nature, I suggested we exaggerate the importance of nature in the raw. We are misled by our aesthetic preconceptions and our sense of guilt about our destructive habits into regarding as natural, and therefore sacrosanct, land that has been actually formed over the years by human intervention.

'Yes. Wherever you turn, except in a very few places where nature has been left undisturbed, like the Grand Canyon, you are dealing with a man-made environment. So the problem becomes, as I see it, not one of conservation but intelligent management.'

I could see that this was true of Europe, which has been worked over for many centuries, but America was surely different? After all, the pioneers little more than a century ago had found a whole new continent to exploit.

'The situation is *exactly* the same in America', he insisted. 'Even the famous prairie that has imprinted itself on American lore was created by man. The pre-agricultural Indians used to burn the forest, which thus became prairie. Then modern man destroyed the prairie and turned it into farm land. It is true that on this continent we still have *some* enormous stretches that have not been modified by man, like the Sierras, the Rockies, and the Grand Canyon, whereas in Europe even Switzerland is a park. My secretary, when she went there, was really shocked to find it so polished. Man has arranged the Alps into a park.'

It was plain that Dubos, much as he respects his professional ecologist colleagues and acquaintances, differs from them on this fundamental point. 'In this country there is a profound cleavage between the professional ecologists and, if I may say so, people like Lewis Mumford and myself. When we speak of the environment we always think of things created by man.' Dubos greatly admires Mumford and has been much influenced by him. 'The greatest man in America', he said, 'a seer'. Certainly they both seem able to involve all aspects of human activity, physical, cultural, and artistic, in their schemes of thought.

In his own profession, Dubos is best known for two breakthroughs arising out of his early work as a medical researcher, one practical, one theoretical. He discovered an enzyme which would attack the microbe that causes pneumonia; and he revolutionized what is known as the Germ Theory by pointing out that disease-producing organisms are not inherently destructive. This was a profoundly significant advance in the way doctors and everyone else thought about disease. Dubos's insight, one of his colleagues told me, 'marked the first time that anyone grasped the point that disease is part of the total harmony'; or, as Dubos himself has put it, in his *Second Thoughts on the Germ Theory:* 'During the first phase of the germ theory the property of virulence was regarded as lying solely within the microbes themselves. Now virulence is coming to be thought of as ecological'.

So Dubos's first concern had been with, as it were, the internal ecology of man?

Yes, of course, because that is my kind of professional activity. I have come to my interest in the environment through medicine and human biology; all my life has been spent in medical research. How did I get going? It was very odd. I worked for two or three years at an agricultural experiment station in New Jersey, went to school at the same time and got a Ph.D. But in the meantime I met Dr Avery, one of the professors here, at what was then the Rockefeller Institute, and he told me about a problem he had. He was working on a microbe that causes pneumonia. His problem was that this microbe is surrounded by a capsule made of a substance somewhat similar in character to cellulose. I told him, "I think one can find in nature another microbe that will decompose the cellulose-like substance that makes the capsule". I started from the premise that no organic matter accumulates in nature; everything is destructible by some other form of life. All one had to do was find the microbe, obtain that enzyme and use it to decompose the capsule. Well, shortly after this conversation I took a job in agriculture in North Dakota, but I had barely arrived there when a telegram came offering me a job here at the Rockefeller Institute to work on my idea. And my work on the pneumonia microbe made me rather famous.'

But how, I asked, did this work get him interested in the environment?

'There were two very different components. I became known for dealing with problems of infectious disease. I tackled those problems at the level of the most orthodox laboratory technique and I was very successful. Then, around 1942, I began working with tuberculosis. While I was a professor at Harvard, my first wife, who was French, developed tuberculosis and eventually died of it. This was what made me interested in tuberculosis. As soon as I began work on tuberculosis I realized that it is a disease which is important or unimportant depending on the environment in which human beings live. My first wife developed, suffered, and died from tuberculosis because she came from an industrial back-

1. Lewis Mumford
Jill Krementz, New York

2. Rene Dubos

3. Kenneth Boulding

ground in France. Also, I think, because when the Second World War came, being French, she was immensely emotionally disturbed by it, which decreased her resistance.'

Apart from his clinical work on tuberculosis, Dubos became absorbed by the history of the disease, 'There is no mention of tuberculosis in the Bible, either in the Old or the New Testament, because those writings dealt with an agrarian people. But if you read the Greek authors of about the same time, Hippocrates and so on, you find them constantly talking about tuberculosis and consumption, because even those early urban environments were very conducive to the disease. So I wrote an account of the whole historical development of tuberculosis, which I called *The White Plague*. In my view I showed quite convincingly that tuberculosis is an important social disease only under certain environmental conditions. It was an immensely important disease in Europe during the industrial revolution; people kept dying of tuberculosis then, whereas in our world it has almost disappeared. Well, this changed my intellectual orientation. From concerning myself with microbial disease *per se,* I began to concern myself with man's response to his environment, and as soon as I started doing that I became more and more interested in the environment *per se*.

'But there was another component in my intellectual evolution, if I may call it that. One of the great laws of physiology is that the internal state of man constantly responds to the external environment. All physiology is based on a study of the internal state. Claude Bernard, the famous French physiologist, developed the theory of the constancy of the internal environment; which was restated later under the term homeostasis. This concept, useful though it was, is not entirely correct. In reality the internal environment is always changing; it never quite makes a perfect adjustment to the external environment. Because of this belief I have also done a good deal of work on, and written about, the forces in the external environment which impinge on man's internal environment.'

Dubos is notably modest about his pioneering. 'I began arguing about fifteen years ago that the kind of environment we are creating now affects health and disease in a different kind of way from how it did in the past, and urging that medical research should be much more directed towards manipulating the environment instead of being concerned simply with treatment of sick human beings. To say I was the first man to do this would be an overstatement. I would rather say I was one of the first. But I was the first, and still, I think, the only person who linked these two concerns together. In my book, *Man Adapting,* I attempt to explain that you cannot define man unless you define the environment in which he functions. I believe this is an invitation for a completely new science of man.'

At this point, Dubos's excitement was almost tangible, as he jumped up from his desk and went to his crowded bookshelves, looking through rows of his own books. 'Now this book', he said, pulling out *So Human an Animal,* which won the Pulitzer Prize in 1968, 'was trying to describe how man's emotions and intellect and culture are conditioned by his environment. As you see, all the time I try to search how the environment affects man and man the environment. Naturally that led me to take a broader and broader interest into the definition of environmental quality. That is my intellectual evolution.'

Apart from his spell at Harvard, Dubos has always been based at the Rockefeller University, although in recent years he has travelled a great deal. 'I am a sort of freelance operator now all over the United States. I spend much of my time giving lectures. When I first started lecturing people were entertained by me, but my ideas had no effect. But I was well-known as a scientist, so I could afford to go on talking about the environment. I knew people weren't taking me very seriously, even though they knew I was a responsible scientist. For the first ten years at least I didn't make any impact. But now – I apologize, it sounds pretentious – they introduce me as if I were the second Lewis Mumford.'

Apart from a house with some land in the country, on the Hudson, Dubos has an apartment just across the street from the University. He has led an exclusively academic life, but in the middle of the city, not on some secluded campus. His observation of and reactions to the least attractive side of New York life fit neatly into another of his most original lines of thought that the greatest threat to mankind from a damaged environment may not be extinction but adaptation.

'I'm terribly disturbed by seeing how in a city like New York, a child, no matter what his social background, is exposed from the first moment he steps outside to noise, dirt, disorder and *ugliness*. I'm sure that this is conditioning people in an awful way. I don't think this kind of thing kills people, that's where I differ profoundly from the more alarmist ecologists, like Paul Ehrlich, and where I fight with Barry Commoner, though we are very close friends. I always tell them that, in my opinion, the danger is not that we will all be killed by pollution and over-population. What is going to happen is that we are going to accept the situation, and make some kind of adjustment to it. We may in the long run suffer physically as well, but the immediate danger is that we are losing our sense of what the environment could and should be.'

Dubos considers that man's capacity to adapt is 'the one attribute that distinguishes most clearly the world of life from the world of inanimate matter.' This attribute is a boon, but also a potential menace. 'The state of adaptedness to the world of today may be incompatible with the world of tomorrow.' It is his insistence on this point that sets Dubos apart from nearly all the environmentalists and ecological activists who have lately caught the public ear. It needed pronouncements of imminent doom, perhaps, to wake up the general public, but Dubos did not lead the chorus. The distinction is subtle, for it is not that Dubos lacks a sense of urgency, but rather that he is asking for a slight, but significant, change of emphasis. And, indeed, as time passes and the ultimate cataclysm seems not very much nearer, his approach appears to be the more balanced and constructive.

What really troubles Dubos is the spiritual deprivation being inflicted so carelessly on so many people. 'There is a young lady working with me here who is immensely sensitive and perceptive, but because she has hardly ever been out of New York she has never seen the Milky Way.' Dubos repeated this grievous detail. 'Never seen the Milky Way! You cannot see it in New York anymore, ever. To me this is a symbol of what really matters. She is not going to be killed by pollution, but she has never seen the Milky Way. And she has never really experienced the fragrance of spring. This is such an impoverishment. I think it is by far the most important influence of the environment on the future of our society. We risk the loss of our sensual perceptions. And if you lose those, naturally you try to compensate by other stimulations, by very loud noise, or very bright lights, or drugs.' Suddenly he was talking, not in abstract terms, but about America's youth. 'They have to create within themselves the sensations that otherwise they might derive from their natural surroundings. There is an immense human problem there. It's not that I am a puritan, far from it. In fact, it's because I am not puritanical that I am so distressed to see people living in a world where they hardly ever have occasion to experience with their senses how satisfying the world – nature – can be.'

I suggested that the new alarm about the environment, and the endless cries of 'wolf' might turn out to be counter-productive – now that people had grasped the general message.

'This is where I differ from Paul Ehrlich and Barry Commoner', he said. 'They are immensely intelligent and fantastically articulate but they have been talking for so long about the end of the world in 1975 that people have begun to treat them as entertainment. Listening to them has a kind of excitement, like going to a big show. So I have decided to write this kind of thing' – here he produced from a pile of papers on his desk a clipping from a recent issue of *Life* magazine which began: 'I am tired of hearing that man is on his way to extinction' – 'I say "No, you probably won't be killed, but look what's happening to the quality of your life".' He looked serious and said, 'I hope that maybe if people realize that their children's lives may be spoiled, this may lead to more effective action. But I'm not very sure.'

What, then, can be done?

'I cannot give a single answer. I doubt whether adults can change their way of life. By the time you are adult you are conditioned by your past, you have made all sorts of commitments, you are almost a prisoner. So even though people may protest against environmental degradation, they won't do anything to change it – which doesn't mean they aren't aware that something is wrong. But I am more optimistic when I consider the young people. The very young, around seventeen, are immensely aware of what is going on. They have not lost interest at all; if anything their interest has continued to increase, and they have become more articulate and better informed. But they do not have the power to act, and the people who do, those who manage the world, will not do anything to change it until we have a disaster. What we need in the United States is the equivalent of the 1952 London smog. In England it was sufficient to kill four thousand people; here we would probably need to kill forty to a hundred thousand before anything got done.'

It was disconcerting to hear Dubos postulate loss of lives so calmly and at first it seemed as if he was contradicting himself. But it is clear from his writings that, although he is sceptical about the approach of a generalized ecological Armageddon, he is sharply aware of particular risks. As befits one of the exponents of the broad view, he is aware that to attack environmental problems piecemeal will not be enough in the long run.

'Of course, there has been an immense change in public awareness in the last few years, but I don't see that it has led to very much. Admittedly there has been some legislation. We are all supposed to change our motor cars but I am sceptical even about that. I have just been in Los Angeles, lecturing to the California Institute of Technology, and when you see that flood of cars for miles around . . . ' Words failed Dubos, and he raised his arms as if to ward off the hordes of vehicles bearing down on him. 'When you see those vehicles, and realize that you can't see the sky above you, then it is hard to believe that it will be enough to put in afterburners or change the petrol or get new engines. We shall have to change something in our social structure, and that is going to be very hard to do. That won't happen until we have a disaster, and sooner or later we'll have one.'

What usually occurs when a scientist becomes a pundit, on the environment or anything else, is that he gradually gives up his original research. It would not be surprising if Dubos had done so, but in spite of his travels, lectures, books, and articles (he was writing a review of Mumford's latest opus when I saw him) he still keeps up his laboratory work. For the past ten years he has been conducting a series of experiments with colonies of mice. 'All the mice possess the same genetic constitution', he explained, 'but we are showing that depending on the conditions of their earliest development, you end up with giant mice, or tiny little mice, skinny animals, or fat animals, with all sorts of different physical characteristics, and even different brain development. That is crucial. It is a demonstration of what I call 'biological Freudianism', to convey my view that all the characteristics of animals and human beings are, if not determined, at least conditioned by their environment during the very first period of life.'

His other research project is, he says, 'much narrower, but very practical. I am doing lots of experiments to demonstrate that the enzymes put into detergents are potentially very dangerous. We are trying to document what those dangers are. The production of these so-called biological detergents is an excellent example of how careless we are in introducing harmful substances into our environment without giving thought to the consequences. If you think as I do about the environment, you must consider specific instances of harm as well as the more general influences which shape all the characteristics of a person. So I still undertake this kind of experiment in the laboratory. What I try to do, in the time I give to laboratory work, is to orient my work towards environmental problems.'

In his work on biological detergents, Dubos brought to bear his special knowledge of enzymes acquired through his work on the pneumonia germ thirty years ago. He had been suspicious from the start about the possible effects of adding enzymes to detergents. He remembered that in 1939 he had extracted two very powerful antibiotics from a strain of bacteria very similar to the one used to produce the enzymes added to detergents. These antibiotics also had drastic effects on red blood cells and the functioning of the kidneys. He wrote to the American Association for the Advancement of Science's Committee on Environmental Alteration outlining these possible side effects of the new detergents.

This warning was ignored at the time, partly, according to Dubos, because it seemed too esoteric, based on largely forgotten early experiments, and partly because scientists were having a hard enough time trying to establish the relatively simple connection between the new detergents and allergies leading to skin disease and respiratory infection. Recently, Dubos has been gathering more evidence about the possible effects of artificial enzymes on the blood. He has found that as well as destroying red blood cells, commercial enzyme preparations can inhibit the growth of bacteria, which could be damaging to health, and can be toxic to the white corpuscles that protect animals and humans from infection. In a series of experiments with mice, he found that a pneumonia infection could be greatly aggravated by exposure to minute amounts of artificial enzymes.

Through a combination of his specialized knowledge and his alertness to environmental matters, Dubos has thus uncovered a new range of alarming possibilities about the effects of the 'biological' detergents on the people who manufacture and use them. The lessons are clear. As he put it: 'There must be an essential philosophy in our society that prohibits the introduction of any technical innovations until one has tested, insofar as one can, the consequences of its introduction both in the environment and upon the human beings who will come in contact with it. This, of course, has not been the case with enzyme detergents. Millions and millions of people are using them, and we have no idea what the result will be. It is criminal to have such a situation. At the very least, it is criminally stupid.'

We left the University together. As we walked, Dr Dubos told me of another preoccupation, the responsibility of the scientist to society. Although Dubos struck me as being by temperament more of a philosopher than an activist, he has been involved for some years in what is known as the 'information movement' among scientists in the United States. As I was later to hear in detail, a group of scientists, led by Professor Commoner in St Louis and by Margaret Mead and Dubos in New York, had started in the late fifties to press for more communication between scientists and the public, so that the public might share in some of the social and political decisions arising out of scientific and technological discoveries. Dubos had recently been discussing this again with Barry Commoner, this time in connection with the attitude of developed to underdeveloped countries. Talking about this he became indignant. 'We lie to the underprivileged part of the world in telling them they can match what we are doing here.' Then he left me at the entrance to his apartment block, a small (by New York standards) and anonymous-looking modern building.

One phrase in particular remained with me after meeting Dubos. He had referred to the attempt to define man in terms of the environment as 'an invitation for a completely new science of man'. I realized that I was in search of a definition of that new science, and an understanding of its components and techniques, if indeed these already existed. Dubos had suggested some possible lines of thought in his latest book, *Reason Awake*. For example:

'The ecological constraints on the growth of the world population and on the production of energy and of goods will generate new kinds of scientific problems.'

'The drastic limitation of family size will probably create social, psychological, physiological, and, perhaps, even genetic disturbances concerning which little, if anything, is known.'

'The distribution and utilization of energy under controlled conditions will require sophisticated knowledge of regional and spaceship ecology.'

'Entirely new technologies, and therefore new kinds of scientific knowledge, will have to be developed to minimize pollution and to recycle the natural resources in short supply.'

'The steady state will thus compel a re-orientation of the scientific and technological enterprise. Indeed, it may generate a scientific renaissance. But this will not happen without a conscious, and probably painful, effort from the scientific community.'

Perhaps, it occurred to me, Dubos mistrusted the tendency of some environmentalists to alarm and pressurize the public because his own main self-appointed task has been to try and arouse his fellow scientists. The key points in his message clearly seemed to be: (1) Remember that it is man we must deal with, not 'nature' in the abstract; (2) Make sure that scientists know what they are doing and that they keep the public informed.

Dubos sees contemporary man poised uneasily between a passive acceptance of scientific technology and a sudden panic rejection of it. Neither of these choices seems to him to be the answer. However, the debate among environmentally-minded scientists, which he described at the beginning of our conversation, is to him a hopeful sign, for only through open discussion can the public be made aware of the vital issues and share in the attempt to control science and technology and use it well, instead of carelessly as in the past. Meanwhile, Dubos will continue to contribute to this process, with practical work like his enzyme experiments, and through propounding his general philosophy with lectures and books.

Kenneth Boulding

THE ARRIVAL OF SPACESHIP EARTH

IT BECAME increasingly obvious, as I travelled around America talking about the environment, that the discussion remained abstract, so much hot air, unless it was firmly spliced into economics. On the simplest level, it was a question of how to pay to clean up pollution, and how to make continuing pollution so prohibitively expensive to would-be polluters that it would not be worthwhile. But soon, more subtle problems arose. How are natural resources to be valued? If it is plain that we will one day have to cease treating the air, the sea, rivers, lakes, and open spaces as dumps, how can we devise a charge for their use? And if, as I had constantly been told, natural resources must henceforward be exploited in a less profligate manner, how are people to be persuaded to consume less, when most economies are geared to persuading them to consume more? Teams of high-powered, environmentally-minded economists are working on problems like these; but of the senior economic theorists, one man in particular, I learned, had taken a special interest in relating ecology to economics, and had thought out a new approach to production and consumption which often lay behind the detailed work being done by others. This was Dr Kenneth Boulding, of the Behavioural Science Institute of the University of Colorado at Boulder. His approach, which has become one of the key concepts for all environmentalists, is known as the Spaceship Earth concept.

Boulding, who is an Englishman, born in Liverpool and educated at Oxford and Chicago, has made his career in America, becoming an American citizen in 1948. Now in his sixties, he still has a noticeable northern accent. He is a tall man, with longish white hair curling down towards the collar, white bushy eyebrows, and the intense blue gaze of a visionary. He is an economist on the grand scale, who has ranged round many problems concerning man as a social animal. He has written about information theory, social dynamics, ethics, and religion (he is a devout Quaker), and has a special interest in the causes of war and conflict. Before he came to Colorado in 1968 he spent fifteen years at Ann Arbor, Michigan, where he was one of the founders of the Centre for Peace Research and Conflict Study.

He is personally much less daunting than these exalted themes might suggest. He seems supremely good humoured, laughs a lot, and avoids highflown philosophical language. He now lives with his large family (his wife is a sociologist from Scandinavia) in a modern split-level house overlooking Boulder in the foothills of the Rocky Mountains. It is a good site for ruminating about the future of man and society; on one side you can gaze out across the seemingly endless plains of middle America, and on the other the snowy mountain peaks rise abruptly to the sky.

We met, however, in more prosaic surroundings, in the ballroom of the Denver Hilton, where Boulding was giving the main address at the annual convention of the Society of Actuaries of America, on 'The Economics of Environmental Deterioration'. He stood on a dais draped with red plush and flanked by the Stars and Stripes, looking like some startled but genial bird of the mountains among the sober-suited, bespectacled, actuaries who sat dutifully taking notes as he spoke. His talk was not at all a textbook analysis, but a series of provocative semi-epigrams designed, I supposed, to get the audience thinking in a new loosened-up way. 'Everything produces both "goods" and "bads",' he told us. 'The goods tend to be paid for, but the bads do not.' 'An automobile is a suit of armour with two hundred horses inside; anyone who has one is a Knight: anyone who doesn't is a peasant.' 'Everyone here is a deteriorating system.' Here the actuaries smiled nervously at each other. I noticed that the man next to me had written in his notebook: 'Pollution is a fashionable problem', and drawn a neat box round it. Boulding continued, 'We all drink somebody else's sewage'. More nervous laughter. 'One way to be rich is not to want anything.' Dr Boulding did not offer this as a serious solution to problems of consumption in contemporary America. His main point was that 'the evaluation of environmental "goods" and "bads" is a political problem'. His message to the actuaries was to urge them to think in hard practical terms about the assumptions underlying our present economic structures and re-evaluate those structures taking the environment into account. It went down well. After several speeches stressing what actuaries as a profession could do to help save the environment 'we are indeed a profession which is making a serious effort' the session ended and Dr Boulding and I drove off to Boulder, some thirty miles away. We went first to his office, a small, ground-floor room on the edge of the campus.

Did his interest in environmental problems arise out of his economic thinking, or had he always been aware of them?

'It goes back, in a very humble way, to before I ever came to this country, when I worked at Edinburgh as an assistant to Lord Astor and Seebohm Rowntree on a study of British agriculture. This got me interested in agricultural policy and agricultural economics, but the ecological thing really began for me with the Princeton Conference of 1955 on Man's Role in Changing the Face of the Earth. I don't know why I was asked to attend, but it got me very interested. Then I took part in the later conference on Future Environments of North America, and the one in Virginia two years ago on Ecological Consequences of International Development. So really I got involved by accident, through being invited to these conferences.'

Boulding, I knew, had become the environmentalists' favourite economist, as his presence at these three key conferences indicated. Why had they chosen him? I asked.

'Well, my interest in ecological *theory* goes back a long way, at least to my book on the reconstruction of economics in 1941. I have a chapter in that on ecological theory. That arose out of an interest in population dynamics; some of my earliest writings as a young man were on population dynamics applied to capital theory. I was interested in the interaction of populations and then, of course, I was interested in general systems. Ecological theory is a very important general system.'

I could see that the general systems approach, which looks for useful patterns not only in different disciplines but in different functions of man and society, could have something in common with ecology in its very broadest sense, but it all seemed highly abstract. What exactly is the relevance of the general systems approach to ecology? I asked.

'Or, the ecological approach to general systems', he said mysteriously. 'Ecology is a very good example of the general systems approach. The General Systems Society started in 1954, when I was at the Behavioural Sciences Centre at Stanford, California. I was one of the founder members along with von Bertalanffy, the Viennese biologist, Rappoport, a mathematical biologist, and Ralph Gerard, a physiologist. Over lunch one day the four of us drew up a manifesto; we defined a general system as a theoretical system that was of interest to more than one discipline. And if you think of ecology as the theory of the interaction of populations of all kinds, this is obviously an enormously important general system, especially if you include evolutionary theory. I think the ecological thing and evolution are really all of a piece, that is, evolution is really ecological succession, or, defined very broadly, evolution is a change in the parameters of ecological systems. You can work out an equillibrium concept here in which all populations are consistent with each other.' Seeing my expression of bewilderment, Dr Boulding added kindly, 'Of course, this is highly formalistic and doesn't really tell you very much. But one of the great problems of ecological and evolutionary theory is that it is a bit contentless. Ecological theory simply says everything depends on everything else, and evolutionary theory is simply the theory of the survival of the surviving. If you say survival of the fittest, you must then ask fittest for what, and the answer is fittest for surviving!'

On one famous occasion, during a discussion after he had given a paper on ecology and economics at the Future Environments of North America conference, Dr Boulding had indulged in a semi-serious outburst of rage, when he rounded on the ecologists and told them: 'You don't have any sophisticated knowledge in ecology. Ecology as far as I can see has been one of the most unsophisticated of the sciences. You are a bunch of bird-watchers. You really are.'

So despite his interest in ecological theory, he clearly would not subscribe to some of the more inflated claims made for ecology as the key discipline for man's future?

'Ecology is still a very soft science', he said. 'It doesn't have much predictive power. It's a set of insights rather than a set of quantitative parameters. We don't really know, from ecology, what will happen if we introduce a new species into a natural system; we just don't have the information. We have a certain amount of information derived from practice. Agriculture, after all, consists of distorting the ecosystem in favour of man. But none of the problems of assessing the ecological consequences of technology has an estimable value. We know ecosystems are terribly complex, and technological intrusions hit us occasionally at crisis points. You have to know the natural history of the organisms concerned, and I'm not sure that at this stage ecology isn't more natural history than it is a quantitative science. There aren't really any "ecologymetrics".'

And the broader implications of ecology just happen to appeal to people who have a certain attitude of mind, who like interrelationships?

'That's right. And that's very like general systems!'

Wanting to get back to economics, I asked Boulding about the Spaceship Earth notion. The vivid simplicity of the image has made it a favourite among environmentalists, and his essay, 'The Economics of the Coming Spaceship Earth', is regularly reprinted in environmental handbooks and anthologies. In it, Boulding argues that we must make the transition from using the earth as if its resources are limitless, and start thinking in terms of recycling precious materials, which means also that we must stop assuming that increased production and consumption are inevitable and desirable. He contrasts the prudent spaceship approach with the carelessness of what he calls the cowboy mentality.

How had he arrived at these ideas?

'It's been a long, slow development. I just don't know when I actually started using the expression Spaceship Earth. I'm not sure Adlai Stevenson wasn't the first to use it at the U.N. in 1965, then Barbara Ward and I started to talk about it, and Bucky Fuller uses it too. I think we all invented it independently; it's an obvious metaphor, and a nice one. I got interested quite early on in the problem of what happens when our resources are gone. Granted that we may have, at present, what I call a linear economy, going from mines to dumps; what happens when either you run out of mines or you run out of space to dump?'

You start having to use your dumps as mines, I suggested.

'Yes, and living on your own excrement. Perhaps you just have to join the two ends of the line together, but you have to get energy into it somewhere, there's no getting around that. On the other hand, the earth is not a completely closed system; we do get energy from the sun. I've tried to visualize an economy say five hundred years from now, when you've had to fall back on using solar energy, but certainly if fusion becomes feasible, this opens up new possibilities. That's the trouble with all these exhaustion problems; resources are a function of knowledge, and we just don't know what the future of knowledge is going to be. In the last two hundred years we've almost certainly been increasing natural resources faster than we've used them up, just by discovering things and developing wholly new technologies. Oil, you see, is only a hundred years old; then there's uranium, aluminium, magnesium, the whole new chemical industry. We are discovering enormous new sources of energy. We can't predict the future of knowledge, so we don't know how worried we ought to be about our resources.'

Here Boulding enquired whether I was familiar with his poem called 'A Conservationist's Lament and The Technologists' Reply'. He told me that he is given to composing verses, especially during earnest conferences: 'I go into a coma and the muse speaks through the coma'. He produced a copy of the poem, written during the Princeton Conference of 1955.

A CONSERVATIONIST'S LAMENT

The world is finite, resources are scarce,
Things are bad and will be worse.
Coal is burned and gas exploded
Forests cut and soils eroded.
Wells are dry and airs polluted
Dust is blowing, trees uprooted.
Oil is going, ores depleted,
Drains receive what is excreted.
Land is sinking, seas are rising,
Man is far too enterprising.
Fire will rage with man to fan it,
Soon we'll have a plundered planet.
People breed like fertile rabbits,
People have disgusting habits.

Moral:
The evolutionary plan
Went astray by evolving man.

THE TECHNOLOGISTS' REPLY

Man's potential is quite terrific
You can't go back to the Neolithic.
The cream is there for us to skim it
Knowledge is power and the sky's the limit.
Every mouth has hands to feed it.
Food is found when people need it.
All we need is found in granite
Once we have the men to plan it.
Yeast and algae give us meat,
Soil is almost obsolete.
Men can grow to pastures greener
Till all the earth is Pasadena.

Moral:
Man's a nuisance, Man's a crackpot.
But only man can hit the jackpot.

These neat verses convey pretty accurately the cheerful, unsentimental, and essentially hopeful tone that characterizes Boulding's approach to the resources problem. Not that he minimizes its urgency: 'At existing rates of utilization, according to Resources for the Future (a highpowered economists' think-tank in Washington) we've got about a hundred years before we need to start worrying seriously. If the rates of utilization increase, if, for example, the poor countries begin to get richer, so that the two billion people in the tropical zones come anywhere near the level of utilization of the temperate zones, this, of course, will increase ecological strain. It's a question of whether increased knowledge can offset the approaching crisis. We have to look at our present situation as a time when we can use our geological capital to create enough knowledge to enable us to do without it. Our geological capital in the form of fossil fuels including oil has to be turned into knowledge. We aren't doing enough about this yet. No one's doing much work on solar energy, for instance. At the present stage it probably isn't worth it, because we've got so much fossil fuel left, but in another hundred years the situation may look very different. There's obviously tremendous leeway with solar energy. Then there's the molecular biology revolution, which may mean we can create new forms of life, artificial algae which will perform photosynthesis much more efficiently than existing plants – that sort of thing. There are lots of possibilities.'

At the same time, Dr Boulding is convinced that to realize these possibilities we must be prepared for a fundamental change in attitudes. 'I've been very interested in the concept of closure, arguing that this represents a very fundamental change in man's image of himself and his environment. This is what I've called the shift from the great plain to the spaceship. Up till now we've really had the feeling that the earth is illimitable and we've just been expanding into it. "A darkling plain where ignorant armies clash by night", as Matthew Arnold said. The idea of closure is comparatively recent; you don't find it before the last ten years. But then it's only in my lifetime that the earth has been fully explored. When I was a boy there were still white spaces on the globe.'

Have most economists now espoused the closure concept?

'Oh no, it hasn't dawned on economists yet. There hasn't been much serious thinking done on it except for a few people with Resources for the Future. My generation of economists was so traumatized and obsessed by unemployment and the depression that this was almost all we could think about. We were the Keynesian generation. Then after the Second World War the great obsession was development, as it still is; certainly the problem of the development of the poor countries is going to be one of the fundamental problems of the next hundred years.'

But what is the point of encouraging development if it simply means getting more countries into the environmental mess that the developed countries are already in?

'Well', he said, looking stern, 'we want to get them out of the poverty and distress they're in now. There's nothing very romantic or pleasant or even very ecological about Mauritius, or Haiti, or India for that matter; these are countries of really desperate poverty. Obviously enormous increases in productivity are required to get them out of destitution. The whole problem is one of joint production of "goods" and "bads". If "goods" are very scarce you put a very high value on them and you're ready to put up with a lot of "bads". As the old Lancashire saying has it, "where there's muck there's money". It's only when you get richer that you become interested in what are really superior "goods", the quality of life, and so on.'

'Of course, you do get an interest in the quality of life at all levels. I was in Japan recently, and Japan has a number of very important lessons for the rest of the world. It was very much like a spaceship, in the Tokogawa period between 1618 and 1868. They cut themselves off from the rest of the world almost entirely, apart from a few Chinese and Dutch merchants in Nagasaki. They really had a spaceship economy and they got on quite well. They reorganized society; they managed to create internal peace; they slowed down the development of technology, agriculture, and the rise of a monetary system. It was like developing in a chrysalis. In 1868 they came out into the modern world and just flew, the only non-European country to do so. I think the long incubation period was very valuable. The history of Japan is fascinating in this regard. They've gone through many of the episodes of Western history in a completely independent way. Now, of course, they're going all gungho for development and hang the pollution. But then Britain did the same; the Liverpool of my childhood was a completely filthy city. It's certainly much better today. It seems to me that the British are in much better shape today, ecologically, than they were when I was a boy. So obviously you can improve these things. The rivers are cleaner, and the atmosphere and the public buildings are cleaned up, and the whole country is so much richer and happier, it seems to me, than it was in the days of the Empire. I'm sure the Japanese will clean up too. It all depends what "goods" you are prepared to sacrifice in order to get rid of the "bads".'

If pollution problems are compartively easy to solve, as Dr Boulding seemed to be saying, to what, I wondered, did he ascribe the recent panic over the environment?

'I'm a bit puzzled by that', he admitted, 'because there's no major ecological crisis on now. There's even been quite a lot of improvement. Los Angeles isn't as bad as it was ten years ago, and look at Pittsburgh; when I first went there in 1933 I got out at the railway station and couldn't see across the street. If you look at the two previous ecological excitements in this country, you see what has been achieved. The first was the one around 1900 which produced the National Forests and the Bureau of Reclamations, which was quite successful ecologically. That was the point when the Americans began to realize they didn't have an infinite country. Then the second big one was the soil conservation scare in the thirties. I remember the dustbowl. I was a student in Chicago in 1934 and the dust piled up three or four inches deep on the side walks like snow. You couldn't go out of the house for days. Kids nowadays talk about ecological crises, but I've seen one. It was horrible, awful, really frightening, with people saying that the Great Plains would have to be abandoned. That crisis produced the Soil Conservation Act of 1936, which has been very successful. The great plains are in much better shape now, because we've found much better methods of farming. Droughts have been as bad since then as in the thirties, but there hasn't been a comparable disaster.

'I think the present ecological crisis is an eschatological crisis: a crisis about the future. Part of it is caused by the approach of the year 2000, which is a magic number; it's a millennial year, so everyone has started to worry about the way things are going. It's an apocalyptic sort of fear. Then there's space exploration, which means we've just really seen the earth for the first time, and it's so fantastically lovely. I don't think anyone had any inkling that the earth would be so beautiful. This experience has contributed to the sense of closure. One of the paradoxes of the space business is that it has produced a sort of claustrophobia. We have realized that even if we go into space a very long way, we've nowhere else to go. The earth is all we've got. That's why the space business is totally different from Columbus. He was discovering a new world: the astronauts have discovered only an old desert. So we get this feeling of lonesomeness. We are turned in on ourselves.'

This seemed a good moment to ask Dr Boulding for his views on the currently fashionable notion that the Judaeo-Christian tradition, which sets man over nature, has a special responsibility for Western man's exploitative attitude to nature. Religion has played a big part in Boulding's thinking; did he think that Christianity should accept a large chunk of the blame for man's environmental mistakes?

'I think that's mostly nonsense', he said. 'Everyone messes up the earth. You don't have to be Presbyterian to start economic development, that's perfectly clear. But, of course, the whole business of "be fruitful and multiply", and having dominion over the earth, this certainly is conducive to an expansionist phase. The Chinese discovered Africa fifty years before the Spanish, but they just thought to hell with it and went back to China, whereas the Spanish and Portuguese went all over the world, once their technology enabled them to do so. I'm sure Christianity had something to do with it. It was expansionist; conceptually it was a world religion. It had a missionary fervour. Whereas it would never occur to a Japanese that anyone else could be a Shinto, which is a religion especially for the Japanese, it would never occur to Christians that Christianity was only for Europeans. But the problems closure raises have to be solved in their own right. It doesn't get you anywhere just to blame Christianity. And, of course, you can always argue the other way, that we need more emphasis on the Christian idea of community and loving your brothers. And the whole idea of stewardship is profoundly Christian; the idea of man as a steward of the earth is Christian and very ecological. I don't know that Eastern religions as such are any more related to the problems of the spaceship. All religions now face problems of coexistence; you can't be quite as much of a missionary on the spaceship as you are on the great plain.'

Curiously, although the economics of the spaceship would seem to require stringent population controls before almost anything else, Boulding has not made population changes one of his central themes. He has merely said several times that the population problem is of major importance. I asked him what he thought of the population crusade launched by Paul Ehrlich.

'The real problem is that we know awfully little about the dynamics of fertility', he said. 'Mortality we understand a bit better, but even there the impact of D.D.T. on mortality in the tropics astonished us. In the U.S. no one expected the population bulge after 1945, but now that's over, and if the decline goes on we could easily get to a reproductive ratio of one – a static population – within a year or two. The same thing has been happening in Europe; it certainly happened in the Communist countries, which are down to very low birth rates. But the tropics face a situation of great potential danger and there could be a real disaster. But they are beginning to catch it. In Mauritius, the birth rate has been reduced very dramatically in the last decade. I'm a lot more optimistic than I was ten years ago. I think the combination of the green revolution, propaganda, the pill, and the coil, is beginning to have an effect. The real trouble is that we haven't got the institutions for population control. My proposal for a "green stamp plan" is a little tongue in cheek, but perfectly feasible.' This plan, which he put forward in a book called *The Meaning of the 20th Century*, (Allen and Unwin, 1965) suggests that 'a system of marketable licences to have children is the only one which will combine the minimum of social control . . . with a maximum of individual liberty and ethical choice. Each girl on approaching maturity would be presented with a certificate which will entitle it's owner to have, say, 2·2 children, or whatever number would ensure a reproductive rate of one. The unit of these certificates might be the "decichild" and accumulation of ten of these units by purchase, inheritance, or gift would permit a woman in maturity to have one legal child.'

'If the pressures were sufficient', Boulding told me, 'we might have to come to something like this. It's feasible; it's just not legitimate. But maybe there are other limiting factors on population growth that we don't really understand at the moment. All population predictions have been wrong, almost without exception, and I don't think the predictions of the sixties are going to be fulfilled. But I really don't know how far the thing will solve itself and how far we're going to have to intervene deliberately. There's a good deal of evidence to show that in America if only wanted children were born, the ratio would fall to one or below. It has already fallen below one in Japan, and they've got quite agitated about it. And this decline is not the result of a deliberate population policy, except in so far as the Japanese have no objection to abortion and never have had, and contraceptives are easily available. The point is that the Japanese are very fond of their children and very ambitious for them. How far people respond to economic incentives, no one knows, but there's not much evidence that the tax system makes a difference. Education does make a difference, however, and opportunities and higher status for women also make a difference.'

Boulding's view is that while our knowledge of population dynamics is so limited, it is counterproductive to scare people; and if the scares lead to defensiveness, and to a desire to stall scientific and social development, then we will be denying one of the more hopeful and positive aspects of our nature. What ecologists must do is gather more basic information.

'We don't really know much about the earth as a total system, even biologically', he said. 'We don't really know what we're doing to the atmosphere, for example. It's the possibility of long, slow, irreversible change that is scary, the kind of effects that may produce a new Ice Age, or the greenhouse effect. After all, we don't know what produced the Ice Age in the past. We don't know how sensitive the earth is. I do think the earth has a great deal of what the systems types call ultrastability: the capacity of one part of the system to take over if another part folds. The earth is complex enough to have a great deal of ultrastability, I'm quite sure of that. On the other hand, there are cliffs, I'm quite sure of that also. One of the great problems is how you find out about cliffs without falling over one. We don't have too much experience of the cliff. What we know is the plain. There are examples in history of irretrievable disaster, on a small scale, like the Mayas in Central America, where a civilization came to an end just like that, on a Wednesday practically, and nobody really knows why. It was probably a revolution that killed off all the people who knew anything. Or there are places that have gone to desert, like Ozymandias' "My name is Ozymandias, King of Kings: look on my works, ye mighty, and despair". 'In the humid zones we seem to be able to get away with near-fatal mistakes; the arid countries are much more precarious.

'Many ecological crises arise because we apply methods which are suitable in one part of the world to another part in which they're not. This is why the tropics are in such trouble today. I've been arguing that science-based technology is pretty much a temperate zone sub-culture. We know very little about tropical ecosystems actually; that was what the conference in Washington two years ago was all about. I must find my poem on ecological awareness . . .'

This long poem is something of a tour de force, and Boulding was evidently proud of it.

Ecological Awareness leads to questioning of goals:
This threatens the performance of some old established roles.
So to raise the human species from the level of subsistence
We have to overcome covert political resistance.
So we should be propagating, without shadow of apology,
A Scientific Discipline of Poleconecology.

Among the very saddest of developmental tales
Is the indestructibility of fluke-infested snails.
Development is fluky when with flukes the blood is crammed,
So the more we dam the rivers, then the sooner we are damned.

Not that Boulding is against development or change. In fact, his main worry about the sudden fashionable interest in ecology is that it may lead to a fear of progress. 'I'm very sceptical of the equilibrium bias of ecologists', he told me. 'The plain fact is that we haven't had an equilibrium state in this particular universe for three-and-a-half billion years, and we aren't going to start now. Life is an evolutionary process. All ecological equilibria on this planet are temporary. They've always been disturbed by evolutionary potential of some sort. The idea that you can reach an equilibrium and stay there is nonsense, especially for a creature like man. I think it's quite likely that biological evolution, as we know it, has been replaced by what I call teleological evolution. The equilibrium concept is an important myth, but it can be dangerous if it diverts people's attention from the necessity of providing dynamic change. I do think that the present level of economic growth has an upper limit which we're not too far from in America – short of some unforeseen technological change.'

Above all, Boulding emphasizes that natural systems have only a limited relevance to the problems of man in society. The difference is that human societies have a teleology as well as an ecology; they have images of the future towards which society may be directed. Thus the movements of the system may be determined by images of the future, rather than by any simple interaction of populations. This non-random element becomes increasingly important with the dawn of social self-consciousness. Nobody in animal society has any image of the future of that whole society, though I suppose when the lion is jumping on the giraffe he has some sort of image of the future which involves having a good meal. But man has an image of the future of the whole universe.

'It's the introduction of teleology which makes the difference between social systems and biological systems. This is why the sociosphere, as I call it, is much more complicated than the biosphere.'

4

Frank Fraser Darling and Charles Elton

THE ECOLOGIST BRANCHES OUT

IN BRITAIN, so far as the general public was concerned, it was as if one man suddenly invented ecology in the autumn of 1969. This was Frank Fraser Darling, and the occasion was that year's Reith Lectures, which he called 'Wilderness and Plenty'. The main achievement of these lectures, which were a wide-ranging account of man's impact on his environment through agriculture, population growth, and technology, was not only to introduce the listeners to ecology as a science, but to impress upon them that it could be a fresh and exciting philosophy of life as well as a branch of biology. The six lectures were strung on a single theme: a plea for a change in man's attitude to nature.

Before the lectures, Fraser Darling was so little-known in Britain that *The Times* described him as an American. After the lectures he was knighted and asked to become a member of a Royal Commission on Environmental Pollution, which was set up by the Labour Government rather sooner than had been intended because of the lectures.

At the time, Fraser Darling was in his middle sixties. He was a leading international ecological pundit, constantly sought after by conference organizers wanting papers, and by foreign governments wanting advice. Yet he had not then, and has not now, ever held a senior post in his own country, either in ecology or in conservation. He was – as one of the main British conservationists, Max Nicholson, put it to me – 'a man of world status hardly known in his own country'.

Part of the reason, I learned, could be ascribed to Fraser Darling's temperament, and the rest to the tenacity of the specialist British academic tradition. And the story of neither

Fraser Darling nor ecology in Britain could be understood, I further discovered, without attention to a brilliant man who, despite the fact that he is unquestionably one of the founding fathers of the science that is now booming, is still as little-known in Britain as Fraser Darling was before the Reith Lectures. This man was the Reader in Ecology at Oxford University, Charles Elton.

It is a measure of the youth of ecology as a subject that Elton, who was virtually Britain's first professional animal ecologist, has only just turned seventy and is still very much a practitioner. His short book *Animal Ecology* (Sidgwick and Jackson, 1927), is still a basic text. Admittedly he was remarkably young when he wrote it (in three months) aged twenty-six. It is not the kind of book that is easily superseded. Indeed, it is a model of its kind. It states general principles, substantiates them with absorbing detail about the sugar-cane froghopper of Trinidad or the bee-eaters of the White Nile, and is written with a precision, lucidity, and wit that make it a rare treat among textbooks. Elton's father, Oliver Elton, was a civilized and distinguished professor of English, and he passed his own talent for writing English prose on to his son.

Charles was a promising Oxford zoologist of twenty-one when he went on a University expedition to Spitsbergen, organized by Julian Huxley, in 1921. It was as a result of that journey, which enabled him to study the workings of a comparatively simple ecosystem, that he began to formulate the ideas which led to his book.

He placed ecology firmly in context: 'Ecology is a new name for a very old subject. It simply means scientific natural history'. Natural history, he went on, had been pushed into the background by zoologists who, after Charles Darwin's revolutionary work on adaptation and evolution, had felt that little more progress could be made without much more classification of animal species. 'Many of the brilliant observations of the older naturalists were rendered practically useless through the insufficient identification of the animals upon which they had worked. Half the zoological world thereupon drifted into museums and spent the next fifty years doing the work of description and classification which was to lay the foundations for the scientific ecology of the twentieth century. The rest of the zoologists retired into laboratories . . . hence the temporary dying down of scientific work on animal ecology.'

It was plant ecologists who showed the way ahead: 'Scientific ecology was first started some thirty years ago (that is, around 1890) by botanists, who finished their classification sooner than the zoologists, because there are fewer species of plants than of animals, and because plants do not rush away when you try to collect them. Animal ecologists have followed the lead of plant ecologists and copied most of their methods, without inventing many new ones of their own.' Elton intended his book to remedy this situation.

Yet although early plant ecologists did much valuable work – the late Sir Arthur Tansley of Cambridge in particular helped to broaden the scope of the subject – their influence became constricting. Fraser Darling once put it like this: 'The early literature of ecology gravely neglected the influence of the biotic factor. Ecology as we knew it fifty years ago was a botanical science primarily, handicapped by a certain restriction of vision associated with those whose eyes are focussed on the sward.'

Elton's book gave a succinct and detailed signposting of the direction animal ecologists should take, the methods they should use ('watch the animal eating . . . examine excretory products . . . examine the contents of the crop or stomach or intestine . . .'), and supplied them with a set of basic principles drawn partly from his own research and partly from a thorough analysis of the research of others. He established the fundamental ideas of population cycles, food chains of varying complexity, and the idea of animals filling niches in the function of conversion of matter. Scientific methods, he said, must be applied to the study of whole communities of animals and their relations with one another. 'When an ecologist says, "There goes a badger",' he wrote, 'he should include in his thoughts some definite idea of the animal's place in the community to which it belongs, just as if he had said, "There goes the vicar".'

Elton was acutely aware that ecology could, and should, be a way of organizing and illuminating existing information. 'The various books and journals of ornithology and entomology are like a row of beehives containing an immense amount of valuable honey which has been stored up in separate cells by the bees that made it. The advantage, and at the same time the difficulty, of ecological work is that it attempts to provide conceptions which can link up into some complete scheme the colossal store of facts about natural history which has accumulated up-to-date in this rather haphazard manner.'

What is most striking in retrospect about Elton's book is that without spelling it out and almost, one feels in the light of his later career, without realizing it, *Animal Ecology* conveyed an awareness that ecologists, if they chose to do so, could extend their subject to cover almost everything. There is a significant phrase in the introduction to the first edition: 'Ecology is a branch of zoology which is perhaps more able to offer immediate practical help to mankind than any of the others, and in the present rather parlous state of civilization it would seem particularly important to include it in the training of young zoologists'. Again, in a passage at the end of the book, he refers to the 'rather uncomfortable feeling' which sometimes comes over an ecologist that 'it might be worth while getting to know a little about geology or the movements of the moon or of a dog's tail, or the psychology of starlings, or any of those apparently specialized or remote subjects which are always turning out to be at the basis of ecological problems encountered in the field'.

But although Elton wrote that 'man is only one animal in a large community of other ones', he was by inclination more of a naturalist than an environmentalist. Having erected his signposts, he turned aside and immersed himself in a study of fluctuating mammal populations. In recent years he has taken up ecological survey work again, with a massively detailed study of the woodlands round Wytham, within bicycling distance of Oxford. Elton, in fact, chose to follow up his own teachings in his own way, on a relatively small scale.

In the introduction to the 1965 edition of his classic, he comments on the changes in ecology since he travelled as a youth to Spitsbergen. 'The subject of animal ecology is still the same, but its methods and ideas have expanded tremendously.' Much more information has been assembled, although still not enough: 'The young ecologist is growing up into a research world of processes, rates of change, controlling factors, patterns, statistical planning and mathematical models and interactions. . . . Scientists have learned a lot more about the variety and composition of different kinds of habitats . . . the conversion of matter and energy by living creatures is being surveyed. Population analysis has become more sophisticated with the use of models and computers.'

It is plain from this new introduction that Elton is fully aware of the scope and responsibility of modern ecology and nowhere does one sense any condemnation of ecologists who take the entire globe rather than Wytham woods as their field of study. 'During this century the activities of man have come more and more to dominate so many situations in the world that it has become difficult, except in purely technological applications of the subject, to separate pure from applied research. . . . Whether in the solution of difficulties about natural resources, carriage of disease, or the quality of human life in the future, animal ecologists have much to give.'

Elton and Fraser Darling are virtually contemporaries, and have been friends for forty years. Fraser Darling met his second wife in Elton's house in Oxford. The two men share a passion for natural history. Yet as ecologists they have taken quite different routes. Partly this is a matter of temperament, which is nowhere better illustrated than by their attitudes to conferences. Fraser Darling loves them; Elton hates them, once refusing to read his presidential address to the British Ecological Society at Oxford. Instead someone else delivered it on his behalf, while he sat in the audience. Elton, as he has got older, has sunk himself more and more in the detailed study of a comparatively minute area. Fraser Darling has ranged more and more widely, generalizing as he goes – though this is not necessarily an activity of which Elton, as his writings show, in the least disapproves.

Fraser Darling's development made it inevitable that he would not find a lifelong home inside the British academic tradition, even though as a young man he briefly held an academic job at the University of Edinburgh. Friend of Elton's though he has long been, the tradition was too narrow for him. It is not irrelevant, perhaps, that both Elton and his younger colleague, Mr H. N. Southern (the author of *Tawny Owls and Their Prey*), preferred not to be interviewed for this book – the only refusals I encountered – which leads me to suspect that in Britain there is a lingering suspicion of amateurs and generalists who wish to intrude upon what is still thought of as strictly academic territory.

Fraser Darling, on the other hand, when I asked if I could talk to him, at once cordially invited me down to his home near Newbury in Berkshire. Fraser Darling's main job these days is as Vice-President of the Conservation Foundation in Washington, DC, which was founded by Fairfield Osborn in 1948 and engages both in research and education. Fraser Darling, for years now, has thus become accustomed to spend half his time in America and half in England, which, however, he regards as his home. Also he is Vice-President of the International Union for Conservation of Nature and Natural Resources which is based in Switzerland. That organization was founded at the UNESCO Conference at Fontainebleau in 1948. Fraser Darling is one of the dwindling group of survivors of that occasion.

He is a large, craggy, rumpled man in his late sixties, who looks less like a philosopher than a gamekeeper who has spent a good part of his life out of doors, as indeed he has. He speaks slowly and deliberately, but with sudden flashes of humour, in a deep voice marked by traces of his Scottish and Yorkshire background. His personal environment is enviable. His eighteenth-century house of rosy red brick is full of books of history, anthropology, and poetry, and he has built up a stunning collection of treasures: early Syrian glass, Chinese porcelain, primitive bronze and clay figures, furniture, and – his special passion – old Central Asian rugs and carpets.

'The real scholar finds the word ecology rather degraded now', he said cheerfully, as we sat in the book-lined morning-room. 'It has become suspect. It's a word that has been taken up by generalists who lack the concentration required to become specialists. Human ecology has never really been able to justify itself. People dropped it and talked about the environment instead. What the ecologist has to look out for is lack of profundity.'

But his own approach was surely that of the generalist?

'I think if I have got anything to contribute it is the broad view. I'm that kind of a mind. There's a virtue in seeing the large, but any scholar could easily say, "He's not frightfully profound, that Fraser Darling fellow". I'm seeing round so many corners and that's why scholastically I'm so damn poor. Of course, you have to specialize before you can be let loose. You have to have your scientific discipline in this field.'

Here, Fraser Darling was speaking with the confidence of a

man who knows that his early research work has a solidity that places him out of reach of would-be detractors. His first work, published in 1932, had what might be thought a limited range for a man who, later, would be recommending a total philosophy for modern man. Its title was *The Biology of the Fleece of the Scottish Mountain Blackface Breed of Sheep* (Berlin, 1932). I asked him how the transition came about.

'I've always been a naturalist', he said. 'I was born that way. It was not just something I took up at the age of fourteen or fifteen. There's no doubt, though, that the greatest influence on my young life was reading Charles Darwin. I was fourteen; I was being prepared for confirmation when I read *The Origin of Species,* and, of course, I just said, "This is the truth; this is it". I went to the minister who was preparing me and said "I can't go on; I have seen a larger truth". Not that I am an atheist, by any means.

'I always knew what I wanted to do in life; but the difficulty was getting through to be able to do it. I could only break into it through agriculture; stock farming was not at all what I wanted, but it was at least a way of continuing to be out of doors and being with animals.' He studied agriculture at Edinburgh University, and did his Ph.D. in genetics. One of his first jobs was on the agricultural staff of Bucks County Council; next, after the Ph.D., he went to work as chief officer of the Imperial Bureau of Animal Genetics where, he recalled, 'I had an immense time to read'. It was then (1928–30) that he read Elton's book *Animal Ecology*. 'It was a big step forward; when I read it, everything just fitted in.'

But he was still feeling his way. He seems to have had an instinct that his kind of ecology must involve the activities of man, but, as he told me, 'In those early days, the human being was still regarded as something far, far beyond the animal.' He began to be interested in population, and in 1932 boldly gave a colloquium on the forthcoming rise in human population. At that time, this was not only unfashionable but contravened all accepted prognostications – for instance, those of J. B. S. Haldane. Fraser Darling recalls the occasion with mock embarrassment. 'It was an absolute flop; I didn't know where to put my face.'

Then in 1933 he hit on an idea for a specific piece of ecological research which was to establish his name as an animal ecologist and lead, eventually, to a massive study of human ecology in Scotland.

'I picked up a copy of *The Listener* on Oban station, saw an announcement about the new Leverhulme Fellowships, and went home and wrote out my idea for a study of the red deer. My wife said: "You've got a proper conceit of yourself, haven't you?" But I knew I was going to get the fellowship. There are these moments in life, you know.'

His two years of work on red deer, which led him to formulate an original theory of the social behaviour of the herd, meant long hours of patient observation of the animals in their habitat, the Western Highlands. During this period of intensive concentration, most of it out of doors, Fraser Darling experienced more keenly something he first felt as a very small boy leaning over a mountain stream: 'The capacity to merge into the natural scene, as if I disappeared and became part of nature, and then gradually crystallized again.' This Wordsworthian sense of identification with nature was an important part of his developing philosophy. For instance, he deeply believes in the importance of saving, untouched, the globe's last few patches of wilderness. While recognizing that most people, through circumstance or temperament, will never have the kind of direct experience he had, he maintains that all of us have a psychological need to know the wilderness exists, even if we never go there. This would explain the peculiar distaste most people feel when they hear of unspoiled country being ruined, or learn that the white beaches of the Arctic are gradually being littered with plastic bags and bottles carried there by ocean currents. When Fraser Darling mentions such things, he evidently feels real pain.

His book, *A Herd Of Red Deer* (Oxford University Press), was published in 1937. The following year he published another ecological classic, *Bird Flocks And The Breeding Cycle* (Cambridge University Press). Armed with a Carnegie research grant, he then went off with his wife and young son to the remote island of North Rona, off the north coast of Scotland, where he worked on the herd of grey seals. Then came the war and life on his base island Tanera, where he worked his small farm and wrote. These books, without being in the least sentimental about nature, are full of emotion as well as detailed observation. He remained marooned on his Scottish island for most of the war, through no conscious decision of his own. Just after war broke out he fell ill, and then broke his leg.

By this time, he had begun to formulate the idea which was to bring together all his early experience of animal ecology, and enable him to bring in people as well. I asked him at which point he realized that he was on to something with the widest possible implications.

'Probably in 1939', he said. 'I was always interested in folk as archeological, anthropological, historical material, but at that time I don't think I loved folk – in the mass, at any rate. It was during the isolation of those island expeditions that I first realized that if ecologists were going to do any good it was no use being a long-haired naturalist all your life. You had to bring your knowledge to bear on life. In 1939, after I'd had a letter from Charles Elton asking me some question about the Western Highlands, I wrote him a long reply describing the devastation, the erosion, and the poverty, and giving my view of the reasons for it. I didn't see the whole picture very clearly at that time, but through my work on the red deer I had begun to understand the consequences of man's impact, especially through tree-felling and the introduction of sheep. I said at the end of the letter, "There's quite a nice human ecological study to be done here". During the war, I kept thinking about what had happened in the Highlands, both historically and biologically. People looked upon the Highlands as if they had always been as they then were, but this wasn't so at all.'

In the Western Highlands, Fraser Darling had found a situation and a landscape where human intervention had been crucial, unacknowledged, and on the whole disastrous. In his book *Natural History in the Highlands and Islands,* published in 1947, he had a chapter on the human factor, which was a summary of his thoughts up to that time.

In 1944 he became Director of a new venture, the West Highland Survey, the first major British study in human ecology. It was to take six years, involve a team of six people, and result in a modern Domesday Book of the Western Highlands, ranging over history, geology, biology, agriculture, and social structure, with details ranging from bracken infestation to lists of available doctors and nurses. 'We spent a lot of time tramping round', he recalls. 'We first tried to find out as much recorded detail as we could. Asking questions is often not good tactics; West Highlanders especially just don't like it.'

Fraser Darling shows no false modesty about the West Highland Survey; when he says it was twenty years ahead of its time, he means it. As a result, although the work is universally recognized now as a landmark, one of the very few comprehensive ecological studies ever published, at the time it had a stormy passage. Fraser Darling's career in Britain, he makes it quite plain, was not boosted by this remarkable piece of work.

In the forties the English political establishment, prodded by the ornithologist and passionate conservationalist, Max Nicholson, was beginning to take scientific conservation seriously. Fraser Darling was at that time firmly identified with Scotland, then even more on the periphery in political terms than now, and had spent many years working in the wilds rather than furthering his career in established academic or bureaucratic circles. His ideas were larger, and his attitude to conservation and ecology more sweeping, than those of most of his colleagues. He was something of an outsider.

The survey was finished in 1950, but it wasn't published till 1955. The Scottish Department of Agriculture weren't too happy about it, and it would not have been published when it was had there not been what Fraser Darling calls 'an inspired leak' of large chunks of material to two Scottish newspapers, the *Glasgow Herald* and the *Stornoway Gazette*.

Meanwhile, Fraser Darling was becoming somewhat frustrated by his dealings with the new British conservation establishment. It was as if they didn't quite know what to do with him. He was not given a job by the Nature Conservancy when it was set up in 1949, and when the U.N. held a conference at Lake Success in America on Conservation of Natural Resources, Fraser Darling was not on the British delegation. However, UNESCO asked him to attend, and he did. He recalls this discomfiture of the English with some glee. 'I was told not to represent myself as a member of the British delegation, which I had no intention of doing, and then I was made vice-president of the conference.'

While he was in America, he was approached by the Rockefeller Foundation, who invited him to spend six months travelling round the country doing whatever he liked. 'Of course, this was absolutely marvellous. Why should the Rockefeller Foundation be interested in this little twerp from Scotland? Somehow or other, I don't know how to this day, they had got hold of a draft of the West Highland

4. Frank Fraser Darling
B.B.C. Copyright

5. Kenneth Mellanby setting a mole-trap at Monks Wood
The Guardian

6. Donald Kuenen

Survey. Their director of natural science research said to me, "This seems like a new idea, we'd like to follow it up". So I went for six months and had a marvellous time. I've been so happy there ever since. My life for the last twenty years has been spent coming to and fro between here and America.'

He found wider horizons and more intellectual scope in America. The conservation movement there grew up without the academic links which locked the movement in England firmly into a framework of universities, literal-minded committees, and government institutions. 'There was a very academic attitude in the British conservation movement in those early days and I never kept my mouth shut. I don't suffer fools gladly.' He recalled with some scorn how when he had insisted, in the early fifties, that the plight of the grey seal needed attention, the Conservancy spent some time arguing that because the seal is technically a marine mammal it was nothing to do with them. He managed to get a closed season for his beloved red deer, but not without treading on a few toes.

He found the conservation scene in America much more exciting. The huge scale of America gave the pioneers as recently as the last century a whole continent to play with. This meant that many mistakes were made; but it also meant that a moment dawned comparatively early – not long after the west had been won – when the disruptions became so obvious that men like John Muir of the Sierra Club and President Theodore Roosevelt realized that American thinking must match the grandeur and generosity of the continent itself if its beauty was not to be ruined and its resources squandered. In Britain, on the other hand, man had, in effect, been managing natural resources for centuries. As Fraser Darling put it: 'We've had a much kinder environment, so we've been slower to realize what needs to be done'.

By the time he made contact with the American environmental movement, it already had an established tradition that linked conservation and ecology, and an intellectually respectable literature that dealt with ecology in the broadest possible sense. Aldo Leopold, Paul Sears, Marston Bates, and Fairfield Osborn had all written about the philosophy of ecology, which they saw as directly relevant to the way man handled nature. Fraser Darling had an immediate affinity with men like these. He recalls with evident pleasure how Fairfield Osborn met him on the dock in New York when he arrived in 1950, took him off to lunch, and invited him to use the Conservation Foundation's offices as his base. It seemed very different from England.

This visit shifted Fraser Darling's centre of gravity from Scotland to America. The West Highland Survey had coincided with a change in his personal life. He and his first wife divorced, and he married Averil Morley, who had been his chief research assistant on the Survey.

'I had to leave Scotland then', Fraser Darling says. 'The Survey was at an end, and the devastation of the Highlands was beginning to tell on me. I couldn't look around me without seeing what had gone wrong. The land was sour and perhaps I was, too. I wanted the flora of a sweet land. My wife knew the chalk country and loved it, and I came to feel the chalk flora was something beautiful in itself. So we moved south.' He has lived in his present house ever since. Tragically his wife died of cancer in 1957, and three years later he married his present wife, a strong, quiet Scotswoman on whom he greatly depends. He has four children, one by his first wife and three by his second, and the house and garden have been fashioned with loving care by them all. But he still goes regularly to the Highlands to feel the wilderness again. The family has a farmhouse overlooking Findhorn Bay and the Moray Firth, with the high hills of the northern Highlands as a backdrop.

In America, Fraser Darling has been specially concerned with Alaska, and with the battle between the conservationists and the oil companies. In 1952 he spent five months there with Starker Leopold (son of the great Aldo Leopold), flying thousands of miles in small planes and 'covering good distances on my flat feet'. They produced, in 1953, *Alaska: An Ecological Reconaissance;* both of them have remained in close touch with the conservation movement in Alaska and with the impact of the great oil strike. Fraser Darling seems to uphold a principle that is characteristic of British conservation; where possible, it is more constructive to talk with industrialists than to attack them.

He played a leading part in the two key ecological conferences held in America since the war, the Princeton Conference of 1955 on 'Man's Role in Changing the Face of the Earth', and the Airlie House conference which he chaired on behalf of the Conservation Foundation in 1965, 'Future Environments of North

America'. 'The Princeton conference was terrific', he said with enthusiasm. 'It was the most intensive intellectual week I'd ever had.' Fraser Darling admits happily to being a conference man. 'I like the talk between times so much. Speaking makes one clarify ideas. I'm not conscious of ever having cerebrated like Rodin's Thinker. The most productive times of my life have been when I've just been sitting around, usually out of doors.'

He became involved, too, in Africa. People were aware of the need to conserve rare and vanishing species of animals but Fraser Darling took a much more radical view. 'From the middle fifties I went to and from Africa for ten years, after what was then the Northern Rhodesia Game Department had asked me over. At a very early stage in that work I had an original vision of truth: the notion of the spectrum of wild life, using the environment, each in its own way. You have to examine what each animal is doing in that habitat, and in which way it is contributing to the life style of the whole community.'

Put crudely, what this means is that it is not much use making efforts to conserve a group of elephants if they eat out their food supply, ruin their habitat, and wreck the habitats of other less conspicuous but possibly also rare, and almost certainly ecologically valuable, animals. It is a measure of the progress in ecological thinking that this idea, which now seems self-evident even to the amateur, was something of a revelation only fifteen years ago to seasoned professionals with years of experience of African game wardenship. Fraser Darling's principles of the spectrum also involved the realization that to separate animals in Africa into domestic, or directly useful, and wild, or only indirectly useful, was a false distinction. He has pointed out how much damage has been done in Africa by the well-meaning introduction, often by colonial governments, of sheep, cattle, and goats and ranching them too exclusively. Wildlife had a definite function in conserving habitat.

His work in Africa led him to state another warning which has subsequently been taken up and stressed by others: the necessity for involving ecologists, or at the very least taking note of ecological principles, in development projects in under-developed countries.

So for more than ten years now, as an ecological pundit much in demand internationally, Fraser Darling has been endlessly travelling (though no longer to the tropics, because of a serious internal operation a few years ago), chairing and addressing conferences, delivering papers, lecturing, writing books. He can be relied on to contribute something of value on an enormous range of topics, from the regeneration of slag heaps in Wales to the importance of nomadic tribes (which he may well illustrate from one of his rugs). Increasingly, in recent years, he has given full rein to his general philosophy. His Reith Lectures are a characteristic sample of his style of thought and expression, combining information with homely wisdom and poetry.

What marks him out from the other philosophically-inclined ecologists is that he is prepared to carry his admonitions into the less sure realm of personal ethics and morality. Although he has no formal religious belief, he has a strong religious streak. He inclines to 'the perennial philosophy', regarding the search for truth by mystics as lying at the centre of all the great religions; and he thinks that man, in all his dealings, should be a giver as well as a taker, in sexual conduct as in his treatment of his environment. His line on the population problem, as outlined in the final Reith Lecture, is characteristic. He asks for a new attitude to human sexuality.

'I believe that thought on the problem of population has been too pragmatic, though understandably so. Any intellectual change might take a long time to filter through society. But such change must carry conviction from within and *a priori,* quite distinct from the hope of benefits to be derived. . . . If our culture learnt the potential quality of human sexuality as part of its very being, it would be more helpful in the world at large, where population control is even more urgent than it is in the West. It is time that both the Church and behaviourism dropped the "animal" connotation, and thought more of the uniquely human potential of a developed sexuality not bound up with reproduction . . . An ethic of sexuality, joined with an ethic of the wholeness of life, giving us a reverence for lowlier forms, and reacting on population growth or limitation, should influence the attitudes of the West towards our exploitation of land and animal life.'

Certain very simple universal truths, by no means only the province of the ecologist, have come more and more to the front of Fraser Darling's thinking; love of place, love of one another, love of children. He ended his Reith Lectures like this: 'There can be no greater moral obligation in the environmental field than to ease out the living space and replace dereliction by beauty. Most people will never know true wilderness, although its existence will not be a matter of indifference to them. The near landscape is valuable and lovable because of its nearness, not something to be disregarded and shrugged off; it is where children are reared and what they take away in their minds to their long future. What ground could be more hallowed?'

Fraser Darling's contribution, then, to the present state of ecology and conservation can scarcely be overestimated. He has pushed forward its sum of knowledge, boosted its international recognition, and widened its philosophical base. On a more mundane level, he links the best of the Elton school and the most eager of the new disciples, and the American and the British traditions. He has stood all his life for the fusion of ecology and conservation, and for the unity of ecology itself. He has always resisted any attempt at separating human ecology from the rest. 'There is only one ecology', as he is fond of saying.

The Leopolds

AN ECOLOGICAL FAMILY

IN ANY enquiry into the origins of an environmental philosophy in America, one name keeps coming up: Aldo Leopold, professional forester, pioneer conservationist, professor of Game Management, and writer, who died in 1948 aged not much over sixty, while fighting a grass fire on a neighbour's farm in Wisconsin. The following year his book *A Sand County Almanac* (Oxford University Press) was published. This volume, a collection of naturalist notes centred on a small property he owned in a remote part of Wisconsin, has been ranked with Thoreau's *Walden* and John Muir's writings about the Sierras. The style of the book is consciously poetic, full of detailed observation and natural lore. In it Aldo Leopold set down the outlines of what he called 'the conservation ethic', a plea for a new morality to govern man's dealings with nature. 'We abuse land because we regard it as a commodity belonging to us', he wrote. 'When we see land as a community to which we belong, we may begin to use it with love and respect. There is no other way for land to survive the impact of mechanized man, nor for us to reap from it the aesthetic harvest it is capable, under science, of contributing to culture. That land is a community is the basic concept of ecology, but that land is to be loved and respected is an extension of ethics.'

Leopold represents a crucial link between the theories of men like Mumford and Dubos and the practical conservation movement. He reached his conservation ethic via a life spent in forestry and wild life management, combined with a born naturalist's passion for observation and appreciation of the beauty and intricacy of nature. He had a profound influence on the attitudes of professional conservationists; it was he who developed the concept of wilderness areas within national parks in America, and who tried to get people to think about the principles involved in setting aside large areas for sport, recreation, and scientific study.

In trying to find out more about this remarkable man, I discovered that he had left a remarkable family behind him, all of whom are still carrying on his tradition and developing his ideas. Leopold himself was of German descent; his grandparents emigrated to America after the Civil War and settled on the banks of the Mississippi. They started a furniture making business, the Leopold Desk Company. Aldo was brought up in Iowa and went to the Yale School of Forestry in the early days of professional forestry when Gifford Pinchot was working under President Theodore Roosevelt to establish sound management policies for the national forests. He married a girl of Spanish origin from New Mexico. Mrs Leopold, now an elderly lady, spends winters on her family property near Sante Fe, but lives most of the year in Madison, Wisconsin, where a group of Aldo's former friends and former neighbours from the university (which has become one of America's liveliest centres of ecological research) have kept intact his country farm and land and bought up several thousand adjoining acres besides as a Leopold Memorial Reserve.

His five children are now spread around the country. Two have university jobs: Starker Leopold, the eldest, is Professor of Wildlife Management at Berkeley, and Carl is a biologist at Purdue, Indiana. Another son, Luna, is a senior hydrologist with the U.S. Geological Survey in Washington. The younger daughter, Estella, is a paleoecologist, studying fossilized pollen grains, with the Geological Survey in Denver, Colorado; and the elder daughter, Nina, was married to a biologist who works in Columbia, Missouri, where she was active in local conservation issues and the birth control movement.

Aldo Leopold dedicated his *Sand County Almanac* to his wife. Much of the book is a detailed record of the country and the creatures that he and his family observed during weekends and holidays at 'The Shack' – the name he gave his abandoned farm in the Wisconsin 'sand counties', a farming area on the edge of what used to be the prairies. Aldo left his whole family with a passion for out-door life.

Estella, a small, dark, friendly woman now in her forties, invited me to spend a day with her and her sister Nina at the personal 'shack' which she bought not long ago in the foothills of the Rocky Mountains, about fifty miles from her home in Denver.

A day in the wilds with two Leopolds would, I felt, put me in touch with a distinctively American tradition. Whereas, in England, ecology and conservation had their roots in the small-scale, loving, but deliberately limited observations of a Gilbert White or a Charles Elton, in America they derived from a large-scale, passionate concern for the vulnerability of a huge continent and a nostalgia for the simple living and close contact with nature of pre-industrial Americans.

Around Denver, the contrast between urban civilization and the natural wilderness is very clear; the incredible sweep of the Rockies rises abruptly from the flat plain. Increasingly, the mountains are being turned into a playground. Aspen, Colorado, is a fashionable ski resort and there are roads to the top of many peaks. But the Rocky Mountains are big: there is still a lot of room.

The sun was hot and the sky an astonishing blue on the day I set off from Denver with Estella and her sister. Although the high peaks were thick with snow, there had only been a light fall on the lower ground beneath the pine trees. After driving for about forty minutes, we turned off at Clear Creek, and took a steep, winding track, with snow banked more thickly on each side. Further on, by now in wild country, we parted from the jeep and set off to walk down to the Leopold land, which is in a valley round the side of a small mountain. There was snow underfoot, enough for Estella and her sister to pick out and recognize the tracks and droppings of various birds and animals, a skill they had learned as children from their father, who could read the snow like a book. (Aldo Leopold, who considered the details of animal behaviour important as well as interesting, had written: 'Every farm is a textbook on animal ecology: woodmanship is the translation of the book'.)

They pointed out deer tracks, trees damaged by porcupines, and an old coyote bone, which they noted with regret. The coyote is dying out in the Rockies they told me, partly because of government persons. They also carefully observed the condition of the ponderosa pines, and drew my attention to places where natural fire damage had burnt the lower part of the trees, which, they explained, is not a bad thing, because the tree can then grow straight and tall, uncluttered by encroaching scrub. The pines were interspersed with aspens, many of them oddly twisted by the strong winds. We saw juniper bushes, yellow pines, and blue spruce.

We plodded on downhill towards the head of their valley. Big rocks jutted out from the hillside, and Estella advised us to be quiet; she had once surprised a deer behind these rocks, and thought we might see him again, but the place was empty except for a few droppings. She showed me a cave where a bear might live. We came to a small stream, its banks criss-crossed with snow-tracks, and the sisters spotted a small brown bird among the rocks in the water, a dipper.

Eventually we arrived at Estella's shack, a wooden single-storey log cabin at the top of a fine open space of rough meadow, with pine forest rising steeply on both sides and, behind it, a magnificent view out across the mountains and a glimpse of white peaks. By this time the November sun was really hot, and the sky still more intensely blue; there was total silence except for the sound of water.

Estella had had the shack for two years, and she and her friends and family were still working on its construction, strengthening the roof and making a huge stone fireplace out of grey slate. She insists on using only hand-carved wood and natural stone, and keeps synthetic, man-produced materials to the absolute minimum. The shack had previously belonged to a man who, she said, loved nature and made a living cutting lumber. This was still gold country, and the stream we could hear had probably been panned for gold.

We sawed logs, made a fire, and lunched outside. Far away we could just see some cows in a distant meadow, a pleasant sight of which Estella, however, disapproved. The pasture was not suited for cows, she explained, because they did considerable damage to the vegetation, which in turn affected the soil.

After lunch, Estella brought out her guitar and sang Spanish songs learned from her brothers. She also sang one she had written herself, a plaintive ballad about an old cottonwood tree. She told me that she could clearly remember, at about the age of twelve, realizing that their weekends at the original family shack in Wisconsin conflicted with the regular church and Sunday school routine; she could recall herself thinking that 'As far as religion went, if priests didn't understand the shack, I didn't understand the Catholic religion'.

Estella's highly specialized professional interest, fossilized pollen grains, seemed remote both from her passion for the wilderness and from her active involvement in Colorado conservation fights. In fact, I discovered that until quite recently she had not been publicly active in conservation at all. 'I was standing round watching everything as a child', she said. 'But my father was quite subtle; he never said, "You kids ought to be interested in all this"; he just used to get so excited himself that we couldn't help getting interested.' She studied botany at Wisconsin, and went on to Yale to take a Ph.D. in plant ecology, studying under Paul Sears, who, as well as being one of the early philosopher-ecologists, had made a special study of pollen.

Studying fossilized pollen grains is a minute, intricate business, which involves finding the grains in rock samples, isolating them, mounting them on slides, photographing them, and studying them as well as their modern analogues under a microscope. By comparing the old pollen grains in an area with newer samples, you can learn a lot about plant evolution or prehistoric vegetation. For instance, Estella told me that the study of pollen grains forty or fifty million years old shows that much of the Rocky Mountains was once covered with tropical plants related to the present flora of South East Asia.

It was only in the early sixties that she found herself taking an active part in a conservation battle. In 1963 her brother Luna told her about the alarming scheme of the Department of the Interior (then under Stewart Udall, who subsequently became a prominent conservationist himself) for a series of dams along the Colorado river, including several in the Grand Canyon. The dams were to be sources of power to pump Colorado River water to central Arizona. 'This seemed like very poor human ecology,' said Estella, 'to plan huge new cities where there weren't any water resources.' So she played a leading part in organizing opposition in Colorado to the scheme, which was eventually defeated, largely as a result of the efforts of groups like hers and the Sierra Club campaign under David Brower, whom Estella greatly admires. Since then, she has remained active locally, and has also, with Luna, become a strong advocate of a more positive attitude to conservation within federal government agencies, such as the Geological Survey and the Forest Service. The two Leopolds make field trips together in the summer, and ask their colleagues and contacts awkward questions. This role obviously appealed to Estella.

Estella obviously admired both her brothers greatly. I went to see Luna Leopold in his Washington D.C. office, which is in a large, imposing, characterless government building. He is a tall, handsome man, expensively dressed (unlike most ecologists I had met), with dark, aquiline features that reminded me of his Spanish ancestry, but an un-Spanish air of never saying anything without thinking hard about it first. When I suggested that for a family with a father like Aldo Leopold, an active interest in conservation must have been inbred, he thought very carefully and then questioned the assumption. 'My father's influence on the five children in my family came very much later than you might have expected', he said. 'It's been my observation that all of us, and we are all now very interested in conservation, came to it pretty late. Even if you have grown up in a household where the ecological approach to nature has been a general philosophy, that in itself isn't enough. Sure, you learn about woodcraft, you learn outdoor manners; but in my case, for example, I was interested in hunting, but not in birds as such, while I was growing up. I thought all that was as dull as could be. I think you have to take a pretty close look at the idea that with the right kind of education people will automatically fall into the right attiude. My own son and daugher are not yet particularly concerned about ecology and conservation, though I think they're going to be, and the background is there when it's ready to be developed. That's the way it was with all of us; and maybe it's the most interesting thing about the Leopold family. The attitude was basically there, because of my father, but each of us has had to develop it through our own experience. You just aren't made a conservationist from birth; you have to fight for it in your own way.'

Luna Leopold approached ecology in what seems at first to have been a roundabout way. He studied engineering, physics and meteorology, and geology. 'My father believed that one day there would be a science of the environment – a science of applied ecology – that would go far beyond plant ecology, which was the best developed branch of ecology at that time. I could see that the kind of thing I was most interested in, which happened to be water and water-related problems, was going to require a completely new kind of education. But my father thought I was crazy to study engineering, and my professors did too. But I was convinced that the only way you can talk to engineers is to be one yourself.'

He joined the Geological Survey in 1950, and wrote a book about flood control in 1953. He began to realize how many government projects were impinging on evironmental values, and to try to work out ways to evaluate the beauty or rarity or remoteness of a stretch of country in mathematical terms. He spent some time at the University of Pennsylvania, where America's leading ecologically-minded landscape architect, Ian McHarg, was a big influence on him. All this, coupled with his passion for working out-of-doors, gradually led him into a commitment to the conservation movement.

As soon as he began talking about his summers in the open air, Luna Leopold lit up. He fetched a couple of big albums, which contained a mass of photographs and what looked like a detailed diary of each day's adventures. 'That's just personal. I've kept a journal all my life.' He pointed out his sister Estella and other members of the family and friends, on various intrepid-looking trips down rivers or up mountains. The Leopold clan, I thought, looked like an advertisement for the ecologically-sound life; handsome, suntanned, and laughing as they canoed down rapids or tramped through forests.

He also showed me pictures of a beautiful lake in Wyoming where he had spent the previous summer with some students, studying the topography and character of the lake. After choosing the area for scientific reasons he got into local conservation problems, especially plans of the Forest Service to build a big camp ground nearby, which would certainly affect the exceptional clarity and purity of the water. 'I intend to make the Forest Service so damned proud of that lake they'll do anything to preserve it', he said.

In the last five or six years he has begun to write and lecture to a much wider public, and at the same time he has carried on a constant quiet battle inside his own and other government agencies. On a small scale, he has tried to get colleagues in his department to acquire some field knowledge of their own: 'I told them to get started on a backyard project. They used to laugh at me, but I said you can learn a great deal about nature and water right in your own backyard, even if you just put a rain gauge in, watch the weather, and measure what happens to the little creek that runs through your property.'

More ambitiously, he has put a lot of time and intellectual energy into working out a precise method for fitting aesthetic and environmental considerations into the calculations on which planning decisions are based. He showed me a paper he wrote in 1969 called 'Quantitative comparison of some aesthetic factors among rivers', describing it as 'A preliminary attempt to quantify some elements of aesthetic appeal while eliminating, as far as possible, value judgements or personal preferences'. He set about this demanding task as follows. First, he went to the Rocky Mountains and toured a scenically dramatic region of Idaho where the Federal Power Commission had been studying an application for a permit to build more dams in Hell's Canyon on the Snake river. Next, he picked out eleven other sites on the region's rivers, and studied all twelve sites in detail. Then he drew up a series of tables. Essentially these brought together in tabular form a whole host of Leopold's factual observations of the physical characteristics of each river site: width, depth, bed material, turbidity, etc., plus 'biological and water quality factors', such as water condition, algae, river fauna, and pollution evidence. He included under the heading 'Human use and interest' facts like number and occurrences of trash and litter per hundred feet of river, vistas, and historic features.

He then allotted each of his thirty-eight factors a score between one and five. Thus, the site at Ketchum on the Big Wood river scored one for erosion of banks, three for water condition, five for river fauna, two for metal trash and litter, three for vistas, and two for historic features. The site on the Little Salmon river scored three for erosion of banks, one for water condition, two for river fauna, one for metal trash and litter, four for vistas, and one for historic features. Leopold then worked out an overall score for each site, and came out, via some sophisticated mathematics, with a graph on which he could plot the relative importance of each site according to a scale of objective values – 'Without any personal preference or bias'. The ones with the highest scores should be the sites most worth saving. When he had finished his study it turned out that Hell's Canyon, where the new dams were proposed, came second. Finally, he used the same methods to evaluate four famous river valleys in existing national parks, and compared these with his test sites in Idaho. The Hell's Canyon site, nearest to the existing dam, scored higher than the sites in three out of the four national park sites, and came a close second to the most famous of all, the Colorado river in the Grand Canyon.

Luna Leopold has applied this ingenious technique to another government responsibility – deciding which bits of forest can be cut for timber. He explained to me that he in fact disagrees with the method of cutting often used (known as clear-cutting, which involves removing all trees, as opposed to selective cutting) but, 'If we're going to do it at all, we might as well do it on a rational basis'. He rated possible areas on a scale of 'technical difficulty' as against 'environmental hazard', and discovered that the Forest Service chose the easier but more environmentally hazardous areas every time. When I asked him why he was working on this rather than someone in the Forest Service, he shrugged and said, 'No one else is really doing this kind of work. Most people seem to think that you just can't put numbers on aesthetics. So I go ahead and try to show them that it can be done, because I feel that we must find new ways of measuring environmental impact, so that we can take that impact into account when choices have to be made.'

He has strong views about the responsibility of public servants. 'Public agencies should re-organize their thinking. The government takes action in the name of the people and I would estimate that no less than half the environmental problems in America today are actually caused by the Federal Government. There are lots of examples. The cross-Florida barge canal, for instance, just one of the many water and river control projects planned by the Corps of Engineers and the Bureau of Reclamation. Then there's the Forest Service and clear-cutting, which in my view leads to serious environmental problems. There's the plan for a new Panama Canal. One proposal was to build it by atomic bombs. The Atomic Energy Commission is promoting the use of atomic explosions to develop underground gas resources. And there's the oil pipeline in Alaska. Now, why is the government doing this kind of thing while at the same time it is talking so much about the environment? What we're beginning to see, in individuals as well as our society, is a sort of schizophrenia which no one knows how to solve. You can't expect governments or society to be consistent, and it's the lack of consistency that makes the problem so troublesome.'

(It struck me in passing that perhaps one reason why the Leopolds need their shacks is that there at least they can lead a simple consistent life, not depending on electricity or plastic mixing bowls.)

Luna Leopold is critical of the existing conservation groups, as well as government agencies. 'The organized conservation groups must take a hard line. Everyone else is compromising: someone has to talk about principles. At the moment, conservationists are simply saying, no, don't put a power plant here, don't put a ski resort there, don't put that road through that wilderness. So the main problem in conservation is still a question of basic attitudes. No one is going to say that better than my father did'.

In 1947 Aldo Leopold had written: 'No important change in ethics was ever accomplished without an internal change in our intellectual emphasis, loyalties, affections and convictions. The proof that conservation has not yet touched these foundations of conduct lies in the fact that philosophy and religion have not yet heard of it. In our attempt to make conservation easy, we have made it trivial.'

But it is the eldest Leopold brother, Starker, who has most plainly followed in his father's footsteps. From the hunting and woodcraft he learned with his father as a child, he progressed fairly smoothly to a career in game management and conservation. Since 1946, he has been Professor of Wildlife Management in the Department of Zoology at Berkeley, where he has a grand office with the feel of an old-fashioned study about it, handsome books on the flora and fauna of America lining the walls, leather armchairs, and pictures of wild duck. He is recognized as one of the senior authorities on wild life in America, and spends a good deal of time these days on government committees, helping to plan national park policies and nature conservacy programmes, in the U.S. and elsewhere. In this, he has taken up where his father stopped: Aldo Leopold's premature death cut short an assignment as adviser on conservation to the United Nations.

Aldo Leopold was a keen hunter all his life. He was the eldest son, and his father was a great hunter: they started hunting together from the time he was able to walk. This pattern was repeated with Starker, a jovial character who was wearing, when I saw him, a red-and-white spotted bow tie. 'I guess my interest in nature and the environment began with he hunting and field trips I went on wih my father,' he said. 'My concern with hunting as a sport maybe led me to feel that the study and care of wild life could be a profession.'

Starker Leopold acknowledges that his father had a tremendous influence on him. 'He was a great man', he said. 'But friends of his have told us that it was quite a while before his philosophical views began to emerge. He was just a cowboy type for a number of years, galloping over the mountains, living with rangers and stockmen. Then in 1912–13 he had a serious illness, and spent a year quietly at home in Iowa getting over it. That changed him. Before his illness, he had been all for opening up the whole country, developing its resources, and putting forestry on a sound economic basis before anything else – ideas he had inherited from Gifford Pinchot and Henry Solon Graves. But after his illness he began to propose within the Forest Service that wilderness areas should be set aside simply and solely for their wilderness value – at the time a bold and not particularly popular idea.'

Like a number of would-be environmentalists, Starker started by studying agriculture, at the University of Wisconsin. 'But I dropped out of school in the early years of the depression and went to work; it was just at the time when Franklin Roosevelt was setting up the federal conservation agencies and I worked for the soil conservation service. Then I went back to school, finished agriculture, then took a Master's degree in forestry at Yale and a Ph.D. in Zoology here at Berkeley.'

So, like Luna, he had studied a combination of subjects?

'Yes', he said, 'interdisciplinary degrees were hardly known so you had to do it yourself.' He then went to work in Missouri for the State Game Commission, studying how to restore the wild turkey and deer to the Ozark Mountains. Like Fraser Darling and the red deer in Scotland, he found that to do this job effectively 'you had to understand the general ecology of the area, including human land use'. Although the emphasis in his own work has been on wild life, Starker Leopold has been convinced from early in his career that ecology doesn't make sense without the human element. He first felt this conviction with full force when he spent some time in Mexico at the end of the war with the late William Vogt, one of the first men to sound the alarm about the impending population explosion. Vogt's book *Road To Survival* was published in 1948 and generally condemned as unnecessarily alarmist. Now he is acknowledged as one of the pioneers of the population crusade. Starker was working with Vogt on a survey of wild life in Mexico, and their discussions about the population problem made him realize that he could not discount human activity and numbers when studying ecology. This was not yet a fashionable viewpoint. He went to Berkeley in 1946, and found that 'I was the only one on this campus who would stand up and defend Bill Vogt and his ideas'.

Among the many ecological surveys Starker Leopold has done since then, he particularly remembers one he made in Alaska with Fraser Darling in 1951, on an assignment from the New York Zoological Society and the Conservation Foundation to look into the disappearance of the caribou. 'Great herds of caribou used to wander over the tundra in their millions: now there were only a few hundred thousand left. What had happened? The government agencies took the view that the native people, the Eskimos and Indians, must have shot them, now that they had access to rifles instead of only bows and arrows; they also maintained that a large population of wolves were cleaning up the remainder. This analysis just didn't make sense to me, in terms of what I knew about deer populations after my work in the Ozarks, and it didn't make sense to Frank either, after his work on the red deer in Scotland. So we went up there for the summer. What we found out was this. In the summer, the caribou migrate northwards over the Brooks Range and out on to the great Arctic plain. There you have miles and miles of tundra, all green and lovely, and when you go up there in summer and see all the caribou wandering around up to their knees in grass, it seems obvious that they aren't suffering from lack of food. But in winter, when most people aren't around to observe what happens, they are forced back over the Brooks Range into the Yukon Basin, and there they winter on the edge of the timber zone, feeding largely on lichens, which grow either on the ground or draped over skinny little spruce trees. The caribou has a special digestive system that enables it to live quite well on this stuff.

'Now, ever since the Klondike gold rush, people have been burning that country to clear it, including the spruce trees and the lichens. And the recovery rate is extremely slow; it takes a century or so for lichen growth to establish itself. Everyone was looking at quite the wrong thing to explain the decline of the caribou. The truth hit us one day flying over one of these big burns; we began to make enquiries and found a man at the Yale Forestry School with some figures. He told us that eighty-five per cent of Central Alaska had been burned during the last fifty years. In other countries people know about managing lichen; the Finns, in Lapland where they have the reindeer, know about it, though in a purely empirical way. No one had made the connection that caribou depend on lichen, until we looked at the total situation, including the behaviour of man. To me, that's ecology.'

6

Monks Wood

AN ECOLOGICAL WORKSHOP

THE ECOLOGICAL scene in America is dominated by individuals; but in Britain it is dominated by a place. Monks Wood, the Nature Conservancy's main research station in Huntingdonshire. There is a larger concentration of ecologists at Monks Wood than anywhere else in Europe, and the work done there, since it was set up in 1960, has provided much of the substance, as opposed to the verbiage, of environmental debate. Monks Wood is respected everywhere – in Europe, because the researches of its scientists form the basis for national conservation policy, and in America because it is firmly plugged into public affairs.

In Britain, professional ecologists on the whole do not, or have not so far, taken to the platforms and television screens to arouse the public about the environmental crisis. They remain firmly anchored either to their university departments or to their jobs in government organizations like the Nature Conservancy or the Department of the Environment. They are, accordingly, a quieter, less dramatic, less eloquent bunch on first encounter than the leading Americans. For the interested layman in particular, the excitement and attraction of the American approach to ecology is precisely that it is relevant to almost all human activities, from economics to sex. Of course there are cautious professional ecologists in America, working in universities or in government-financed institutes like the Patuxent Wildlife Research Centre in Maryland. But on the whole American ecologists are an excitable and exciting bunch. The approach of the British is much more sober.

No one at Monks Wood spends much of his official time speculating about the global future or the metaphysics of ecology. Most of those who work there seem if anything to be disconcerted rather than gratified at the way ecologists have shot up the scale of public interest, suddenly promoted from being an obscure group of earnest butterfly-collectors and bird-watchers to being the potential saviours of modern man. Their job, or 'remit' as they often call it, is modestly entitled 'conservation research': the management and protection of land and wild life.

They are in a sense civil servants. The Nature Conservancy was set up by the Attlee government in 1949, and has been under the Natural Environment Research Council since 1965, so that everyone at Monks Wood is employed by the government. Most of them feel this arrangement helps them to get things done by enabling them to apply pressure from within, but one could argue that their equivocal position makes it harder for a real fuss to be kicked up when necessary. They are supposed to advise and warn, and certainly the machinery exists for them to do so. Both government and industry have numerous joint committees and special groups intended to ensure that the research results coming out of Monks Wood do not remain of academic interest only.

Monks Wood is a group of low modern buildings, made of dark ochre brick and glass, seven miles outside Huntingdon on the edge of the flat fen country of East Anglia. It takes its name from the beautiful and ancient wood alongside. It is like a miniature new university set down in the fields, and the men who work there seem a subtle English blend of academic and civil servant; clever but not flashy, amiable and articulate, innately cautious, and given to grey flannel trousers, checked woollen shirts, and tweed ties.

In contrast to many of their American counterparts, the ecologists at Monks Wood, almost to a man, would rather discuss the details of their work than the general principles involved. A number of them feel strongly that the new and intense public pressure on ecologists is a mixed blessing. They all dislike the dramatic, sensational treatment of what, to them, are intricate, subtle subjects. Over lunch in their bright, modern canteen, where they sit informally at long narrow tables eating substantial but unexciting meals of mince and steamed puddings, you are likely to hear some irritated muttering at the latest extravagance of the media's amateur environmentalists.

Once I was there the day after a television news item in which a woman in Arizona had told horrific stories about the effects of defoliants on animals and people. 'Terrible, wasn't it?' said one man. 'So overstated. I thought she was rather enjoying the publicity.' When I asked him how he felt about the ecologist's own new publicity, he said, 'Having worked in the wilderness for so long, it came as a bit of a shock. On the whole, it's a good thing, I suppose, but it can also be dangerous, because it may encourage decisions to be made on insufficient evidence.'

Another luncher, Dr J. Dempster, who has spent five years investigating the effects of D.D.T. on the small cabbage white butterfly and the brussels sprout, said he felt particularly anxious about the dangers to science that could be caused by public pressure for results. 'If we as scientists get our names linked with extremists, it doesn't do us any good', he said fiercely. 'Our scientific reputation is the dearest thing we have. Our responsibility is to point out the evidence.'

All present were very conscious of the sheer amount of work there is to be done. Ian Prestt, who specializes in birds of prey (which are particularly significant in the study of pesticides, as they come at the top of the food chain), and now works for the Ministry of the Environment in London, said ruefully: 'Francis Crick says you'll never get a Nobel Prizewinner in ecology. There are too many variables, and it requires very extensive work for long periods of time, at the end of which you have only unravelled one corner of the situation'. 'The trouble is', added Dempster, 'that man is only really interested in what is going to affect man'.

Dempster, a slight dark man with the nervous intensity of the researcher who loves the work for its own sake, is a member of the Toxic Chemicals and Wild Life team. Before joining Monks Wood in 1964, he had been a lecturer in insect ecology at Imperial College, London. 'I've been looking at the side effects that the application of insecticides to a crop produces on the animals which inhabit the crop', he said. 'I've concentrated on D.D.T. and the cabbage white. We know that D.D.T. persists for a very long time in the soil, and many of the predators of the pest, like the harvest spider and the ground beetles, spend much of their time in the surface layers of the soil. But the pest itself lives on new leaves which are free from insecticide. So if you spray with D.D.T. this year, you greatly increase your chances of a cabbage white problem next year, simply through the amount of D.D.T. left in the soil. This kind of thing has led me to conclude that in many situations you only aggravate the problem by the use of a persistent insecticide.'

It was as impressive to hear this detached, factual account by

Dempster as to listen to the most impassioned attack on D.D.T. imaginable. 'It's got me interested in alternative ways of controlling pest species', he said. 'Within certain limits, I think you can alter the ecology of your crop in such a way as to reduce the chance of having to use a chemical. Many of the predators I'm working on are dependent on a ground cover of vegetation in which they live during the day; they then come out and attack the rest at night. So, by manipulating the ground cover, and making sure there's plenty of weed present, you will increase your predators and knock back your pest; but the snag then is that the weeds will be competition for the crop so you get a reduced yield anyway. As an alternative, we've tried undersowing the crop with clover – which incidentally puts nitrogen back into the soil – and in our experiments this worked well; it encouraged predators, kept the pest down, yet allowed the crop to grow, so that we got an increased yield but no pest problem. Now the question is: is this method agriculturally practicable on a large scale? To investigate further, we've handed over to the School of Agriculture at Cambridge. So you see', he said, with what sounded like a note of relief, 'my research is not very relevant to pollution on a world scale'.

But it was relevant to the search for biological control of insects, I suggested.

'I prefer the term *ecological* control', he said. 'This permits you to manipulate any of the various factors: timing of application of chemicals, or when you sow the crop, as well as the classical idea of biological control like introducing a new predator. Anyway, I'm sure that here in Britain progress is more likely to come through manipulating the way you grow the crop, rather than through introducing a specific creature. But there's rather little of this work being done here.'

Dempster is a living proof that you need not be a crusader in order to be a good ecologist. 'You get strong personalities who jump on any bandwagon to make a name for themselves, basically', he said. 'I've no doubt that there are perfectly good ecologists making statements every day to the public, but I think there is a certain reluctance by the better ecologists to stick their necks out. Once you start making public statements on public issues, you stop being as critical of the evidence as you should be. A research worker needs to be cut off a little from the public limelight.'

But surely, I said, if ecologists have insights and information the rest of us lack, they have a duty to share them, even if the subtleties of their research get blunted in the process.

Dempster looked doubtful. 'It's arguable that an enlightened guess by an ecologist is worth more than an enlightened guess by a non-ecologist', he replied, 'so perhaps it's as well that some people are willing to make guesses. But a fair number of them are going to be wrong. And all your training tells you not to go beyond your evidence.'

The Toxic Chemicals and Wild Life division at Monks Wood, to which Dempster belongs, is the largest research group there, and the bulk of its work over the past five or six years has concerned the effects of persistent organochlorine insecticides on birds, including the peregrine, the sparrowhawk, the kestrel, the barn owl, the wood-pigeon, the buzzard, the golden eagle, the heron, the carrion crow, the song thrush, and the bengalese finch.

Walking through the offices and laboratories, one is surrounded by crates and boxes of eggs to be tested, eggshells to be weighed, and jars and plastic bags containing different parts of birds' innards to be analysed; and there are several large refrigerators full of frozen bird corpses. Whenever some new disaster affects bird life in Britain, the evidence – these sad specimens – finishes up at Monks Wood. The search for clues and answers to the questions where, why, and of what these birds died is like a long, slow, detective story, but there is seldom, if ever, a straighforward answer or a dramatic denouement.

The Irish Sea bird disaster is an apt example. For some time before the discovery of thousands of dead birds in the autumn of 1969, the division had been collecting evidence of the presence in birds of a range of organochlorine chemicals used in industrial processes, called PCBs (Polychlorinated Biphenyls). The scientists were disturbed by the ubiquity of these chemical traces, conjecturing that they might be having important effects on marine and bird life. When quite large amounts of PCBs were found in the dead birds brought in from the Irish Sea, particularly in guillemots which were the worst hit, the temptation was to blame PCBs immediately. One of the ornithologists working on the problem, John Parslow, showed me guillemots packed in the freezer on my first visit to Monks Wood in the spring of 1970. Over a year later, I asked him how the investigation had progressed. He was just off to the north of Scotland, still in pursuit of further evidence, even though the Irish Sea episode was officially closed and a full report published.

'What I'm working on now is a follow-up on the PCBs', he said. 'We had been looking at organochlorine residues in seabirds, meaning the levels in their eggs and so on, since 1962, and we were on to the presence of PCBs three or four years ago. Then we found high levels of PCBs in the Irish Sea birds, which rather surprised us, but we weren't really able to conclude that PCBs were the cause of death. There were unusual gales, a food shortage, the birds were in moult, and they were all very thin. It may not have been PCBs alone, or PCBs at all; it may have been a natural disaster. Nevertheless, the episode showed that the Irish Sea had a lot of PCBs in it. I'm following up the story by looking at guillemots and their subsequent breeding success in the contaminated area, and then comparing them with a control group up in Scotland, where we know that PCB levels are fairly low.' (Guillemots nest on very steep cliffs, so Parslow was getting his climbing gear together.)

So the guilt of PCBs was still unproven?

'It's going to be very hard to show whether they are affecting these birds or not', he said. 'I think myself that they probably aren't, but its difficult to prove a negative.'

One of the biggest companies producing PCBs, which are used in large quantities for making plastics and insulating material, is Monsanto Chemicals, which had withdrawn PCB products in Britain while the research continued – a gesture which pleased the Monks Wood team greatly. 'We will continue to monitor the PCBs over the next few years and it will be very interesting to see what happens to them in the environment', said Parslow. 'We don't quite know how they get there, and we don't quite know their effect. We're also looking at metals, mercury, lead, cadmium, and arsenic: the birds from the Irish Sea showed quite heavy levels of some of these things.'

Something that worries the experts is the hidden longterm effect these substances may be having on small forms of marine life. If a dramatic and flagrant event like the death of thousands of sea birds is so hard to explain, it is infinitely more difficult to predict – let alone to forestall – what may be happening to fish, plankton, and all the valuable crops of the sea on which man will almost certainly depend, before long, to help meet the world's growing appetite. Two years after the Irish Sea incident, the American Oceanographic Institute at Woods Hole, Massachusetts, found that PCB levels throughout the Atlantic were alarmingly high.

As well as birds of prey, sea birds, badgers, and occasionally bats, the division studies the effects of new agricultural techniques. For some years they have been engaged on a special study of hedgerows, which are disappearing from Britain at the rate of about 6,000 miles a year – a serious matter for wild life because it reduces the number of possible habitats. In Lincolnshire, it was discovered that really old hedges, usually parish boundaries, dating back to Anglo-Saxon times, contained nine different species of plants per acre, whereas hedges 200 years old have only two or three. The division have also been studying birds and insects in hedges. They examine roadside verges in the same way, observing the effects of car exhausts, weed killers, and bulldozers on the vegetation, flowers, animals, and insects. Anyone whose dreary journey along some motorway has been suddenly cheered by a patch of poppies or willowherb will appreciate at least the aesthetic value of this work.

Because the Toxic Chemicals and Wildlife division is the largest, and its world the most varied, the other two main divisions at Monks Wood, the Lowland Grassland and Grass Heath team, and the Woodland Management team, tend to be somewhat overshadowed, but their modest-sounding work is every bit as crucial. The Grassland and Heath team investigates the management and conservation of lowland grass areas, which means, after all, most of southern England. They are particularly interested in what happens to chalk grassland when it is grazed by sheep and cattle, which means that the vanishing wild flowers, or 'arable weeds' as they are now ominously called by the experts, are under thorough investigation. The most dramatic single fact I learned at Monks Wood was that the flowery English meadow is on the way out; everywhere it is being replaced by homogeneous green fields of highly-productive, specially-selected grasses.

The Lowland Grassland and Grass Heath division includes a historical geographer, John Sheail, who works on the history of land management, and who has spent several years investigating, and has written a book about, the ecological impact of the rabbit.

The rabbit, Sheail told me, has a dramatic history. It was introduced to Britain by the Normans in the twelfth century, and cultivated for its fur and meat. Gradually the rabbit went wild, but spread slowly, probably, according to Sheail, because its natural predators were more numerous than they are now and winter food was scarce. Then during the nineteenth century the rabbit population exploded, and by 1950 there were some sixty million rabbits in Britain. In 1953, the myxamatosis virus, which had been introduced deliberately into Australia to stop the rabbit scourge there, spread to England from the Continent, and before long rabbit mortality was around 90 per cent, reducing the rabbit population to its early medieval level.

Sheail, a thin, pale young man, believes passionately that historical study of animals and of land use is an important part of ecology. 'Past land use is often the reason for present-day botanical variations', he told me. The botanists, in particular, are very conscious of the recent decline of rabbits, for they used to keep large tracts of grassland nibbled short, thus allowing flowers to thrive and preventing the scrub from growing up too quickly. Sheail has helped to set up a Historical Ecology society within the Conservancy, and wants ecologists to work more closely with archeologists and local historians.

The Woodland Management division, headed by an articulate and healthy-looking tree expert named Richard Steele, has two main aims; first, to advise owners and managers on how to look after their woodlands, and, second, to research into woodland ecology and conservation. Steele feels strongly that woodland conservation must be undertaken in an especially practical fashion, since woods are economically important and trees are bound to be looked on as money. The division has been conducting a national survey of trees native to Britain, including topics like 'acorn production and fall'; 'pattern of tree distribution and the girth and distance relationship between trees and their neighbours'; 'squirrel damage to nuts'; and 'the ecology of the wood ant'.

Some of this research is carried out in the wood right next door to Monks Wood, though the scientists' activities, like those of the wood ant, are not easy to spot, as I discovered one sunny spring day when most of the Monks Wood staff were spending their lunch hour strolling in the bluebells. Only one huge solitary oak – the wood is mostly hazel – showed where they were at work; it was surrounded with planks and platforms, because someone was surveying all the fauna that commonly inhabit oak trees; here and there, a suction insect trap was visible through the leaves.

None of the people I met at Monks Wood were calm or complacent about environmental problems. On the contrary, they have frequently led the way in diagnozing and drawing attention to new dangers. But all of them, without exception, still believe in working through the existing system, and getting on with their own particular job.

7

Kenneth Mellanby

THE EXTREME ENGLISH MODERATE

AS HEAD of Monks Wood, Dr Kenneth Mellanby could be described as the apotheosis of the British school of ecologists. He is a distinguished scientist, a senior administrator, and an accepted member of the professional and government establishment. Apart from his early entomological research work, his main achievement has been the building up and organizing of the Monks Wood team over the last ten years. He is also the author of a short but important book addressed to the general reader, *Pesticides and Pollution* (Collins, 1969), in which he takes a calm and balanced look at all Britain's pollution problems.

Mellanby is a neat, greying man in his early sixties, with a constant look of amusement, quick movements, and a brisk way of speaking. He divides his time between Monks Wood, meetings in London, and lecturing round Britain. He lives with his wife and undergraduate son in an old and beautiful seventeenth-century farmhouse, in the Huntingdon fields, where they are making a rambling garden out of what used to be the farmyard. As befits an ecologically-minded family, they are leaving much of the wild cow parsley, and even some healthy patches of nettles. 'If you kill all the nettles, you won't have any tortoise-shell butterflies', he told me.

I asked him what he thought of the new interest in environmental matters.

'There's a tendency for people to say everything's getting worse', he said. 'The trouble is that some ecologists, and some of the people who like to be called ecologists, but who are really publicists without any particular knowledge of the subject, feel they can get a better hearing by saying everything is getting worse all the time. I think it's much more sensible to say – where this country is concerned at any rate – that despite our increasing population, and despite our increasing industry, in certain limited fields there has been a remarkable improvement. Also the general public has an idea that a lot of things aren't well but they don't quite know what is wrong and there's an awful tendency for them to blame the wrong things. For instance, people are getting excited about pollution from the motor car, which in Britain probably isn't terribly serious.'

I said it did no harm, surely, to draw attention to the pollution threat from cars, but Mellanby disagreed.

'There's a risk I think, that the government will feel it must legislate against the car and its effects on air pollution, which will cost money and may not make very much difference to the environment. Whereas the same amount of money spent on things which really are known to be damaging would get far better value.'

So where should the main effort be made?

'I don't think there's any doubt that the most serious pollution in this country is in water, from industry and insufficiently treated sewage. Even when conditions have been improved, as the Thames has been – there's no doubt that the Thames is cleaner than it's been for over a hundred years – from the ecologist's point of view it's still polluted. We may be able to drink it – we are pretty coarse sort of creatures, we can drink fairly filthy water – but the game fish, like trout, can't stand it. However, some improvement has taken place even in the worst rivers and everyone knows roughly what we have to do. This is where we should be making the effort.'

Mellanby, like some of his staff, obviously deeply disliked the indiscriminate alarmism which surrounds the ecology craze in America. I asked him whether there is any basic difference in approach between American and British ecologists.

'I think', he said, 'that the real work being done by scientists has a very similar approach; but they have this rather high-powered propaganda machine as well. The revivalist technique is very short-term. When this generation of American students have settled down in their suburban homes polluting away like everyone else, the next generation may be concerned with some quite different issue.'

Mellanby became interested in the environment via medical entomology, insects and health, a specifically useful branch of biological science, which perhaps explains his matter-of-fact attitudes. He, too, had a strong interest in nature as a child. 'I used to go out and collect flowers with my mother in Scotland at a very tender age', he told me. 'I started off at Cambridge as a medical student but I got diverted. I decided at that time that I wanted to be a botanist, and after I had done part one of the Tripos I spent the summer of 1928 collecting wild flowers in the Canadian Arctic, which was a very interesting experience, and one of the things that was impressed on me was the terrific pressure of insects. We had to wear veils and we used the rather inefficient repellents we had then, but even so, you could run your hand down your arm and get a handful of squashed mosquitoes.' Dr Mellanby's face took on an expression of quizzical disgust as he recalled those long dead insects. Deciding to specialize in insect psychology, Mellanby went to the School of Tropical Medicine in London, and travelled to Africa to work on the tsetse fly, and Finland to study low temperature mosquitoes. 'It was ecological work, really,' he said. 'I was looking at insects in relation to climate.'

'Then came the war. There was a tremendous outcry at the beginning of the war when children from the towns were evacuated to the country; everyone complained that they were crawling with lice. But the records said that only something like one or two per cent were lousy. I got a grant from the Ministry to travel round the country and investigate and I found that the official figures were absolute nonsense. In cities like Liverpool, 70 per cent of the girls had lice. But the medical officers kept saying that they were not really worried about lice; they were worried about scabies, the itch. They were certain this was on the increase, and they felt sure that in wartime conditions, with bombing and everything, it could be a terrible problem. So I said we'd better look into it.' Mellanby plainly much enjoyed recalling this particular problem of human and insect ecology. 'It's rather interesting, what happened. It shows the way scientific planning is often done. One thinks that high-powered committees of experts plan research, but with this problem I had a harebrained scheme and got a lot of pacifists to volunteer as human guineapigs and we tried to infect them in various ways. We found that almost everything people thought they knew about scabies was wrong. Everyone believed that the way it spread was through blankets and bedding. We found that bedding and blankets were of no importance at all, and saved a lot of money by showing that it was unnecessary to fumigate bedding. We also found that whereas one application of a suitable medicament killed off all the parasites, the itching symptoms very often continued; this was because the medical people were applying the wrong kind of ointment, which made the sufferers itch even more, and was nothing to do with the original pest. By investigating the parasitology and ecology of the insect we succeeded in stopping all this. It really worked out very well. I reckon since then we must have saved three or four million pounds a year', he said. It was clearly a source of some pride to Mellanby to think that he had thus simultaneously reduced human discomfort and saved the community money.

Mellanby went on to study scrub typhus in South East Asia, and went back to the School of Tropical Medicine at London University after the war as Reader in Entomology. 'I only stayed two years,' he said, 'because I went off to start the new University at Ibadan, in Nigeria'. (He wrote a book about this experience, *The Birth of Nigeria's University* – Methuen, 1958 – and acquired a taste for peanut soup, which he makes himself.) 'I had six or seven years in West Africa, mainly as an academic administrator, though I did a little work on mosquitoes. Then I returned to the School of Tropical Medicine, to get back into respectable research from immoral administration. Then I went for six years to the Agricultural Research station at Rothamstead, where I was head of the entomological department; this took me into a new field of agriculture. Then ten years ago I came to the Nature Conservancy, as head of Monks Wood.'

Thus Mellanby's background, until the early sixties, had ensured that he was thoroughly acquainted with the beneficial effects of modern chemistry in dealing with organisms that no one can feel very sentimental about insects that thrive on dirt and cause disease and crop failure. 'When I was in the army', he said, 'I was very much concerned with the introduction of organochlorine insecticides like D.D.T. I have probably been more responsible than most ecologists for recommending the widespread use of pesticides. We had to use them under military conditions. And I started work in tropical medicine, so I was concerned with medical problems; and if you consider that saving human life is important, I don't think there's any doubt that D.D.T. is the greatest boon that mankind has ever had. Anyone who asserts that we must immediately stop using D.D.T. in under-developed countries would be responsible for instant death on a scale that would make the numbers of people killed during the war in concentration camps negligible.'

Mellanby sounded quite fierce as he made this dramatic attack on the opponents of D.D.T. Nowhere is the judicious attitude which he adopts to all environmental hazards more evident than in relation to D.D.T., which has come to be regarded by some reputable scientists in America and Europe, especially Scandinavia, as ecological enemy number one.

In his account of D.D.T. in *Pesticides and Pollution,* Mellanby writes: 'I well remember when I first heard what was then a closely guarded secret about this new "wonder chemical". It seemed miraculous, for it killed insects at dilutions which, at that time, seemed greater than could easily be explained, yet it seemed practically harmless to man. . . . In general the impression was that we had at last discovered the perfect insecticide, which was quite safe to use if reasonable precautions were observed. Under wartime conditions it really was very nearly perfect.' Then in a typical aside, he goes on to say: 'It may be significant to note that the acute toxicity of D.D.T. is almost exactly the same as that of the drug aspirin . . . the same quantity (about an ounce) of D.D.T. would be lethal to most men'. He then explains that there is 'one very important difference'. Aspirin is not retained in the body, i.e. the effect of a number of small doses is not cumulative. But D.D.T. is. When D.D.T. was first widely used, this fact was unknown and inevitably it took time for the cumulative effects to become apparent.

Mellanby draws a clear distinction between the use of pesticides in places like Britain and Europe, where insect pest problems are relatively small-scale and easy to handle, and the needs of places like America and the developing world. This is not a distinction that is universally accepted; and it is hard to see, in theory, why any country should continue using substances that other countries have rejected as ecologically damaging. But in practice there may not be any alternative, and Mellanby, at any rate, knows where he stands. 'There is at present nothing as efficient as D.D.T. for purposes of malaria control in under-developed countries. We know this isn't going to last for ever, as we are getting an increasing number of resistant varieties of insect, but so far this isn't very widespread and D.D.T. is still the best method we have. But it is obviously going to be causing a good deal of ecological damage. You've got to balance the two things together. Now we know, in Britain, that we can get all the control we need without using substances that have really damaging effects; and this is only going to cost us about a million pounds a year extra. So we can afford to do it, like the Swedes, who have tended to lead the way in banning certain chemicals. They were hardly using them at all anyway. But if we are going to ban or restrict the use of these chemicals in under-developed countries, then we shall have to give them an awful lot more money and more expert advice. Eventually, I think, most of these chemicals will be phased out. But at the moment, people who give the impression that wicked industry is selling these poisonous chemicals to under-developed countries because they can't sell them in Europe are talking nonsense. It's nonsense even financially. Most of the time firms make a loss on D.D.T. because it's a terribly cheap chemical. And the patent ran out long ago. If I were a chemical manufacturer I'd far rather be selling small quantities of some expensive chemical and making an enormous profit than supplying huge quantities of D.D.T. very cheaply.'

In the British context, that Mellanby could thus put himself in the shoes of a chemical manufacturer does not seem odd; but it would sound very peculiar to nearly all ecologists and conservationists in America, where for ten years or more battlelines have been drawn between those who support and those who oppose chemical pesticides. To Mellanby and most of his colleagues, this confrontation is senseless and destructive. 'Of course', he said, 'the Americans have special problems, but they do rather go to extremes. They either want to ban the stuff completely or they want to blanket spray their crops with thirty times the usual dose. There's this enormous polarization. You get the ecologists condemning pesticide use, and the industry taking the view it did when *Silent Spring* was published; they tried to denigrate Rachel Carson in every possible way, writing the most frightful things. The American chemical industry got together a big publication on D.D.T. recently, and it had references to "the hirelings of conservation", and that sort of stuff.' Mellanby's whole demeanour expressed mingled amusement and disgust at this crude and abusive approach. 'Well, we don't get that kind of thing here, partly, I think, because a number of our industrialists are keen naturalists in their spare time, which helps. I know one insecticide factory that used to cause a lot of trouble, polluting rivers and so forth. A year or two ago they appointed a new managing director who is a very keen fisherman, and he has seen to it that his factory has been a model over pollution ever since. We are fortunate in having a substantial number of keen conservationists in the chemical industry. Everything is far from perfect, obviously, but the collaboration we get between industry and conservation is very good. We frequently have parties of executives from chemical companies coming here to be indoctrinated. There's no doubt that these companies are trying hard to find better chemicals; the trouble is that the very properties of persistence now objected to in these organochlorines made them exactly what we were looking for twenty years ago.'

It is hardly surprising that this apparently simple and optimistic approach to pesticide pollution should seem, at worst, infuriatingly smug and at best wilfully innocent to the more embattled American ecologists. At an international conference not long after our conversation, Mellanby provided a caricature of his own attitude when he told a large international audience that his recipe for solving chemical pollution was to 'give every managing director of a chemical company a fishing rod'. Mellanby is fully aware, of course, that the situation in America is on a totally different scale to that in Britain or Europe, but his general approach is conditioned by his deep conviction that these problems can all be solved by a combination of intelligent research, common sense, and good will. He looked totally baffled when I told him of the view, widely held in radical circles in America, that big companies are deliberately fostering the ecology boom so that they can cash in on the need for antidotes to their own products, thus improving their profits and their public relations at one blow. 'What does it matter what their motives are?' he asked. 'They are delivering the goods!' Obviously, for Mellanby, ecology is not a tool with which to alter the social structure.

Despite his support for the use of chemicals in some circumstances, Mellanby, like most other modern biologists, is convinced that a combination of chemical and biological methods will ultimately be the only effective and safe technique. But, characteristically, he is moderate in his appraisal.

'All integrated control really means is common sense', he said. 'Everyone who has to pay money to buy pesticides would much rather buy less and rely on natural enemies; this is going to happen more and more. But it means an awful lot of knowledge. With biological control, you are up against the weather, for example, which you can't foretell, and you do sometimes get an unexpected increase in some insect populations because of an extraneous factor. So there has been a tendency to play safe and use chemicals. What we are most against, here at Monks Wood, is the prophylactic use of pesticides, that is, using them just in case something may happen. We need more research so that we can say when you can expect an attack on a particular plant. We do this to some extent already, for example, with wheat bulb fly. We do egg counts in the soil in the autumn, so that you can tell whether you ought to protect the soil against the parasite next spring. This is the sort of thing farmers need to know; but it's not easy.'

Most of the famous and spectacular successes of pure biological control happened some time ago. The cottony cushion scale insect in California was ruining the cotton crop in the 1880s so an Australian ladybird was introduced to deal with it – successfully, until insecticide sprays directed at quite different insects killed large numbers of ladybirds. In Australia, the prickly pear cactus, introduced from America, was becoming a serious weed so an American caterpillar was brought in to deal with it. Perhaps the most successful recent example of large-scale biological control was the introduction of myxamatosis to Australia in the fifties. Mellanby rather enjoys pointing out the emotional inconsistencies connected with pest control. Although farmers in Australia, and later in England and Europe, benefited enormously from the virtual disappearance of the rabbit, many people felt there was something cruel and disgusting about fostering the disease deliberately.

'People are rather disappointed that recent research hasn't been as successful as some of those early efforts, which happened without nearly so much scientific investigation and fuss. Those discoveries were really just a good guess by a man who probably was also a good ecologist and knew the creatures themselves. Our modern scientists who work the whole thing out on a computer ought to be terribly successful, but so far they haven't been, apart from some quite striking successes on sugar cane pests in the West Indies. It may be that we've already tackled the easiest and most dramatic situations and are now left with pests that are too complicated for simple methods.'

Mellanby also has shrewd and pertinent comments to make about the inconsistencies that arise when people get emotional about different ways of dealing with pests: 'People now tend to take the line that chemical control is ethically wrong, and biological control is something wonderful. From this position, they then interpret biological control very widely. People were very thrilled with the idea of the sterile male technique'. This is a method that involves breeding large numbers of insects in captivity, and subjecting them to radiation which renders them sterile. They are then released at the mating season, and infertile eggs result, so the pest is drastically reduced. The technique worked with the screw worm in America; but Mellanby is not too impressed. 'Chemosterilants are really just another sort of insecticide', he said. 'And they may be a carcinogen or have other unpleasant side effects which could be harmful to man. Also, though the screw-worm technique was a very wonderful piece of work, it was terribly costly; the budget has been many times the cost of all entomological work done in Britain. We've all spent a lot of time trying to think of other instances where this technique might be used. I think, in the end, it may be used to mop up the last survivors of a particular pest . . . but then this may not be altogether a good thing. A lot of us are very worried about the wisdom of trying to eradicate certain pests completely. Is this the right thing to do? Or are you going to upset the whole ecological balance and maybe get yourself into worse trouble?'

There is a special Director's Laboratory at Monks Wood, which accomodates studies that don't fit in neatly elsewhere, and Mellanby uses it to follow up a particular interest of his own: moles. 'It's curious', he said, looking tremendously pleased, 'that moles interest the public more than most creatures. I nearly always have a mole or two about the place. My house, Hill Farm, is usually referred to around here, I regret to say, as Mole Hill Farm.' He started taking a keen interest in moles comparatively recently, and has written a delightful and erudite book on them.

How was he drawn to study the mole?

'It was rather an interesting reason', he said. 'You know we've got this International Biological Programme going – an attempt by biologists all over the world to co-ordinate their research and activities; well, one of the things it's interested in is the potential biological productivity of the soil. Now if you read the accepted textbooks you find that the mole is said to be tremendously voracious. It apparently eats twice its own weight of earthworms per day, and we know that you only get this sort of earthworm population in a very fertile soil. So I said to myself: you see mole-hills in the most unlikely places, therefore your moles may be a convenient shorthand. It would be nice to have some simple method to indicate whether areas which are not producing much in the way of crops might perhaps be able to do so after all. So I started looking into it, and I soon found that every statement that is commonly made about moles is wrong. First of all, moles are very much less voracious, and can subsist on much less, than we thought. So you find moles in very infertile areas. But the essential thing about the economy of the mole, which must have been very well-known but had never been written about, is this: a mole doesn't dig anything like as much as you might think. A mole makes an underground tunnel system, which is its feeding ground; it then feeds on what comes into the tunnel. So if you look at a fertile area, you only find short tunnels. These may exist for years, and you get very few mole hills. Here at Monks Wood, we've got a good mole population which lives in this style. But in poor soil, like the Breck, you get tremendous mole activity; the poor creatures have a very hard life because there's so little food that they're always having to burrow, and they make lots of mole hills. So you almost get an inverse correlation between mole traces and soil fertility – but not quite. So in the end, I'm sorry to say, the mole made *no* contribution to the International Biological Programme.'

Mellanby has done much work on the habits and physiology of moles: 'The mole lives most of its life in underground tunnels, through which it lollops along at up to two-and-a-half miles an hour. This is, considering the small size of the animal and the narrowness of the tunnels, a substantial speed, and were it constantly to blunder into the walls it would damage itself, particularly it's very sensitive nose which would receive the brunt of the encounter. . . . Moles have been observed pursuing and catching such agile prey as frogs. . . . There are reports of plant food, including cabbage, being eaten, but I have never succeeded in persuading a mole to try such substances'. Mellanby has also collected literary references to moles: 'John Clare was very good', he said approvingly. 'The two authors who observed moles best were Clare and D. H. Lawrence.'

One fact is clear: the mole is not a friendly, cuddly little animal, as represented by writers like Kenneth Grahame in *The Wind in the Willows*. 'They are very solitary, and they hate each other', Mellanby said. 'Ordinarily, when they meet, two moles will fight, under certain circumstances to the death; they have a brief truce during the breeding season. Their territorial behaviour is very interesting. There's a good deal of controversy between certain authors over territoriality. Lack, working on birds, has given the impression that the food supply determines density. But Professor Wynne Edwards disagrees. With the mole, you could argue it either way. In an area with plenty of food, you only get two or three moles to the acre and they stay in their burrows, so they only colonize a small area, and make negligible inroads into the amount of food available. In an area of pasture or deciduous woodland there would be enough food to keep fifty moles or more, but you never get anything like that number. Something is controlling the numbers, but we don't know what. It's a very stable population, especially in woodlands. Of course, woodland was the original habitat of the mole, because all southern Britain was covered by woods. But in poorer soil, I'm sure the populations are being completely controlled by the amount of food. So it appears that in some parts of the country the Wynne Edwards theory is being supported, in others the Lack hypothesis. This must be true of a great many other animals, but the mole shows it rather neatly.'

Apart from their intrinsic interest, do moles have much ecological significance?

'In grassland, they have a very important ecological effect. Dr Alec Watts has discovered that it is mole activity that affects the whole succession of plants in the Breck areas. They throw up open soil which is then invaded by plants.'

Another Mellanby interest is the spontaneous regeneration of oak seedlings on railway cuttings, which he watches from the train window on his frequent trips up to London. This kind of unpretentious study is very much part of the English naturalist tradition, and Mellanby rates it highly. 'People sometimes dismiss ecology as "glorified natural history",' he said. 'Well, what's wrong with that? I think natural history is a very good thing.'

8

Donald Kuenen

A EUROPEAN EXPERIMENT

HOLLAND IS one of the smallest and richest countries in Europe; it is also the most densely populated, and the investment per head in nature conservation is higher than in any other European country. In 1968, this investment was 0.22 per cent, compared with 0.02 per cent per person in Britain, and the figure is especially significant because the Dutch spend the money on scientific wildlife research and ecological studies rather than on saving the landscape, which is where most of the money goes in other European countries. The Dutch have one of the best-known conservation research institutes in the world, the Dutch State Institute for Nature Conservation Research, and although the staff number only about forty, their calibre has made the Institute a leading force in Europe.

One of the outstanding Dutch ecologists is Professor Donald J. Kuenen, formerly Professor of Biology at Leyden University and now director of the State Institute for Nature Conservation Research. I had heard about him from Dr Norman Moore of Monks Wood, who told me he thought Kuenen one of the best and most effective ecologists in Europe, combining academic brilliance with a practical grip of public affairs. Kuenen is a tall, elegant man in his late fifties, well-tailored, with a quizzical, ironic manner and the polished, authoritative air of a top civil servant.

It had occurred to me that the Dutch, because of the geography of their small, flat, low-lying country, much of it below sea-level, must have had centuries of experience of land management. Perhaps this explained their position as the most conservation-conscious European country? Maybe the Dutch by now are natural conservationists? Dr Kuenen looked amused but sceptical when I suggested this.

'I do not really think so, though you may be right', he said courteously, in perfect English. 'Our situation is much more complicated than would appear from your suggestion. We have grown up with the idea that Holland has too much water and suddenly we've had to realize that there isn't so much of it that is usable. Of course, when you get a break in a dyke there is too much, but that is an accident. Apart from a few incidents, the low-lying parts of the country have indeed been very well handled, and the consequence has been that the technologists, especially the water technologists, have set their stamp on the whole Dutch set-up.'

So, in fact, because of a largely successful tradition of land and water management, it is extra hard for the Dutch to take environmental problems seriously?

'We are very much used to feeling that if we want to alter the natural conditions in our country, we can just go ahead and do so. There is great confidence in the quality of our engineers. So, in the west and north especially, we rely much more on technology than on conservation. But because the west and north have always been extremely fertile and well-cultivated, the less fertile, sandy areas in the south and south-east have been looked on as comparatively unimportant. Population growth was much lower, industry didn't develop, the standard of living was much lower, too; and for three or four centuries agriculture in the south has been very backward. So in that part of Holland there developed an aesthetically-beautiful landscape in which these backward agricultural practices fitted beautifully. There were small fields, lots of hedgerows, and no great concentrations of population. It was an extremely stable community culturally, as well as in its landscape and agriculture'.

So that, paradoxically, it was in the least well-managed part of Holland that conservation habits developed?

'Yes. But in the last fifty or seventy years, as our overall population has grown and our agriculture has become gradually intensified over the whole country, the countryside is being altered there too.'

I enquired whether Dr Kuenen himself was a cradle conservationist or a convert, and whether, like so many of the ecologists I had talked to, he had been a keen child naturalist.

'Oh yes', he said. 'I was brought up in the country, in the flat part of Holland, and for years I had an aquarium. I must have started when I was about eight. At first I had two jars, then three, then four, then came the unbelievable day when I finally had a glass aquarium. And I've had one more or less ever since.'

Was his main interest then in aquatic creatures?

'Yes, but it was really accidental; it could have been flowers and insects just as easily as water bugs and weeds. I just happened to have friends who were particularly interested in aquaria. I'm a great believer in the accidental in life.'

How did his interest in water bugs lead into ecology?

'I studied biology at university, at Leyden in the thirties, and in Holland you have to study all zoology and botany, without specializing, for the first three years. We did have one hour a week on something called ecology, which was in fact just glorified natural history and not much more. The emphasis was entirely on anatomy and physiology, and systematics of course. Then I did my doctoral thesis on some hydro-biological problems, concerning water plants and insects, but when I'd finished there was no job in that field. So I took a job with an industrial firm, in entomology, studying the chemical control of insects.' (Kuenen was one of the select group of senior ecologists and conservationists, like Mellanby, and, on a smaller scale, Commoner, with experience of being as it were on the wrong end of the pesticide business.) 'When the war started, the firm cut down on staff, and in 1941 I switched over to a government research team. The fundamental difference between this job and the one in industry was that in industry you began with the chemical and said "now, let's see how we can use this", whereas on the research team we started with a particular problem, and then tried to work out not so much which insecticide to use as when to use it, and how to adapt chemical control to the biology of the creature. I was involved in this work up until 1950, when I went back to the University of Leyden – by that time acutely aware that the intense application of insecticides like D.D.T. led to the flare-up of other insects. That was how I became interested in biological control as an alternative to pesticides.'

What particular problems had he worked on at this time?

'Fruit growing. We saw how the use of insecticides and fungicides affected the intricate relationships between plants and the animals living on them. We also became aware of the effects of chemicals on the soil, and the influence of fertilizers on the growth of certain pest insects. All this led me towards acceptance of the ecological approach to insect control. When I went back to the university in 1950 as a full-time professor my job was called General Zoology, but the emphasis was very much on ecology and applied ecology. I consider myself to have been teaching ecology for twenty years. I came to subscribe completely to the integrated control approach to pest control. I still remember how we younger scientists were worried about the whole question of the genetic resistance of insects to insecticides, and how the older entomologists couldn't keep up. I shall always remember that experience as a warning to myself. I remember how we used to laugh at the old fellows who simply couldn't believe it when something that they hadn't learned about when they were young appeared before them. They simply refused to recognize the existence of the problem.'

So how did this worry link up with conservation?

'Well, it soon became plain, of course, that a change in insect-control methods was important not only for conservation reasons but as a hard necessity. Then in 1959 I was invited to prepare a paper for an I.U.C.N. (International Union for the Conservation of Nature) conference in Warsaw on the side-effects of chemical control – the ecological implications. That was, in fact, the essence of the problem in which people like Dr Moore in Britain are specialists now: What is happening to the environment as a result of insecticides? Is it possible to find other methods which will help to protect nature? That was how I became associated with I.U.C.N., which I have been involved with ever since.'

But surely he must have been by instinct a conservationist before that time?

'I have always been emotionally a conservationist', he said looking thoughtful and severe. 'This admission raises a fundamental point. So many people try to rationalize their emotional feelings in an attempt to make their commitment acceptable to others. I have done this frequently, but at least I know that I'm doing it. I think a lot of people don't realize how much emotion, how much irrationality, there is in this whole conservation business. In part, it should be accepted as such; although there is also a rational background to it, this is still incomplete and very often ill-defined. For example, there was talk in Holland of digging up a large chalk hill area in the south to make a cement works. This would have meant destroying a great number of caves in the hillside which bats used for hibernation in winter. People protested that this was an attack on nature; then they began to say that it must be bad for agriculture, because if you killed the bats which feed on insects you would get more insects and less crops. Theoretically, a sound approach, quantitively, a negligible effect; but symbolically very important.'

What Kuenen was getting at here seemed to me highly significant. We are not yet prepared to allow a small but symbolic example of man's damaging effect on nature to carry the weight in argument that it deserves. We are still hung up on the need to prove that such actions do calculable damage, not just to one small part of the natural community, but to man himself. This tends to mean that we do not take environmental damage seriously except when it occurs on a huge scale. It is, of course, no use trying to argue that the obliteration of the caves of a few bats will affect the life of every Dutchman; but this does not mean that such destructive acts are unimportant. For one thing, they are important cumulatively.

'It's not just the bats that are going out', said Kuenen sternly. 'Numbers of hedgehogs are killed each week on the roads in Holland, and they obviously must amount to a certain percentage of the hedgehog population each year. I don't know the exact figure but we are doing some work on it; traffic certainly is affecting the hedgehog, there's no doubt about that. Also, as people take better care of their gardens, there's no room for the hedgehog. Therefore it is on the way out. Moles are going out as well, and so are lots of other insectivores. Each of these effects is practically negligible, but if you add them up you get a disruption of the biological system.

'Here is a difficulty. As biologists we have been trained to see the full complexity of the ecosystem, and to observe the implications of anything we do to nature; but we have too little quantitative data to explain these things to people who are not prepared to believe them already. It is very difficult to convince people just by telling them that isolated incidents all add up to something serious, when we can't even say for sure at what rate the hedgehog population is declining. To know for sure, you would have to have one man studying the problem for, say, five years; it is all long-term, and very complicated, and the diversity is too great. If you study hedgehogs, you know nothing about moles. If you study moles, you know nothing about shrews, etc., etc. Each problem has got to be studied separately.'

Dr Kuenen became extremely animated as he contemplated the astounding, infuriating variety of the natural kingdom. 'Let me get on to one of my hobby horses', he said. 'In general, when a technologist, a chemist or an engineer, makes something, he knows exactly how it works. He is never confronted with a mysterious, complex mechanism; either he has made it himself or one of his friends knows all about it. But the biologist or ecologist is confronted with something that he knows he'll never fully understand. He is therefore forced to use vague interpretations. Obviously he uses as much quantitative data as he can discover, but he knows he'll never get it all. So his whole approach to his subject is necessarily much less exact than the technocrat's, not because he is a fool, but because his field of study is so much more complicated. Therefore he is despised by the technologists, who say: "The man doesn't know what he is talking about"; while the biologist can only say to the technologist: "Some things are too complicated for you to handle your way".' Kuenen subsided in exasperation.

Obviously this built-in clash between technologists and biologists had bedevilled his own work, and affected the handling of environmental problems in Holland, though it is just as relevant elsewhere. I asked him to outline his work on pesticides.

'At the time when I used to do my own research, which, alas, I don't do any more – I direct institutes and sit on committees – my work was concentrated, as I say, on integrated pest control, and especially on the possibility of using mites to control red spiders in orchards. All my work was related to agriculture, because with the very high productivity we need in Holland we have an intensive, high-yield agriculture and there has to be very precise insect control. I first realized the problem of resistance, as well as the effect of insecticides on the stabilization process generally, through my work on the red spider mite. We saw it coming up quite early after the war, though, of course, the first real case of insect resistance was in California in 1906. But the crunch came in 1946, when Geigy, the people who had developed and introduced D.D.T., said they were finding resistance. This led to an explosion of work on the problem which has been going on ever since. We saw resistance develop with the red spider; we've seen it since with aphids; and it has started to happen with mosquitos in Africa. It will happen wherever there is systematic control with a particular insecticide. The terror is that even if you then switch to another chemical, resistance will build up even more quickly. We shall have to stop using insecticides in this way within ten or twenty years. Therefore, it is misleading to say that we can go on using them in places where they are really necessary. In a little while, they simply won't work, and meanwhile we shall have been pouring these chemicals into the environment and thus creating other problems. What we really need to do is to press on much faster with biological control methods, in order to keep up with a situation which we can already see is deteriorating.'

I asked him how the Dutch were using pesticides, since the scare.

'People have retreated from excessive use of these chemicals, because of the warnings of the biologists', he said. 'For example, before autumn and spring sowings of barley, wheat, and so on, farmers have to apply a fungicide. They used to add an insecticide, just to be on the safe side, but that has been stopped. All spray programmes are being reduced as much as possible. Official advice used to be that "to spray four times is always safer than to spray three times", but now it's the other way round. There's been a very clear change in attitude over the last fifteen years or so towards using as little as possible, rather than as much as you can afford. We have banned dieldrin and aldrin, though only in the last two years; we can't get our Ministry to impose bans until the situation is really very bad.'

I mentioned the great debate between the Americans and the British about the best way to tackle the pesticides problem: the Americans advocting an all-out attack on the producers and users, invoking legal measures; the British propounding the virtues of voluntary agreements and as much friendly contact as possible. I asked Kuenen where his sympathies lay.

'It has become clearer than ever to me recently that to adopt the American approach in Holland, or in Europe in general, would just be going absolutely the wrong way. We already operate, to some extent, the system advocated by the British. The biologists have frequent open discussion with farmers and farmers' organizations. We are also trying to build up contacts with producers.

There are a few angry young men in Holland who want to start the kind of legal environmental defence organization that the Americans are setting up. I am convinced that for us this idea would be fundamentally wrong. We need more contacts with industry, not a pitched battle. I think in this the British have pointed the right direction' – here Kuenen raised his eyebrows and added in tones of some irony – 'as they so often do.'

Apart from pesticides, what, I asked, were the major environmental worries facing the well-organized Dutch?

'Well, our pollution problem from industry and the towns is something incredible. There are only enough sewage works in Holland to cope with about thirty per cent of the sewage; the rest goes straight into our canals, rivers, and lakes. Industry has improved a lot, however. There have been two or three cases of really bad pollution from factories lately, and in each case when they were told to stop it they immediately did so. They don't have to be convinced; for the last five years or so they have been very alert. Our plants build high stacks, for instance, to release sulphur dioxide fumes higher up, so that the stuff is considerably diluted by the time it comes down to earth. Of course, the poor Swedes think it is carried by the winds and comes down on their earth, and I think they are probably right. I'm sure they are right about the unusual acidity of their lakes – they wouldn't dream that up – and it's hard to see where else it comes from except Holland, Britain and Germany.'

Because of their geographical position, the Dutch are particularly conscious of the international nature of European environmental problems. Above all, they are on the receiving end of the Rhine, central Europe's major river and still its main industrial drain. Thus the Dutch have a keen interest in the pollution habits of their neighbours.

'The condition of the Rhine is extremely important', Kuenen said seriously, 'and I don't know how it's going to work out. The International Rhine Commission (which has been in existence since 1950, representing Switzerland, France, Luxembourg, Germany, and Holland) has still only got engineers on it – no biologists, no environmental people.'

But surely ecologists should be part of this crucial group?

'The Commission experts feel they know all about water so why pull in biologists? In a way I see their point. What would a biologist do except kick up a row? The problem is already obvious. Of course, the technologists only talk about microbial aspects and bacteria, they don't know anything about fish.' For the first time in our talk, Kuenen seemed a little despondent, then his crisp optimism reasserted itself. 'But I think one day soon they'll start to clean it up properly, in spite of everything.'

Remembering the great Rhine fish disaster of 1969, when several millions of fish died in three days, while Europe nervously waited to find out how far the killer chemical had spread, I asked Kuenen about the effect of this disaster on public opinion in Holland where it must have been a major trauma.

'Yes, it was', he agreed, 'but it had no practical results. Everybody got frightfully excited, but the Germans systematically tried to obstruct the investigations. You can find all this in the newspapers of that time; in the first place they didn't warn us, although they knew what was going to happen because they saw the dead fish floating in the middle Rhine. But they said nothing. So that it was only when the dead fish came floating into Holland a day or two later that we were alerted. Then the Germans said the stuff must have dropped off a ship, and they led the investigators astray looking for ships which had arrived in Holland. That took up another few days and it was simply a red herring. Finally the Dutch themselves went up the Rhine and made the measurements.' Dr Kuenen bristled with rage as he remembered this episode. 'Even then the Germans pretended that the chemicals must have been washed off the hills by rain. Finally, our Secretary for Public Health had to go to the government in Bonn and say: "Here are the measurements, this is how it was, this is what we've found, this is where it came from, here's the factory that makes the stuff" – Kuenen was banging the desk in fury by this time – "and here is my report". So then, finally, they had to confess, and they said, yes, possibly some of the stuff had been spilt by accident. They must have known that from the very beginning. It was very bad. Of course, we have very good relations with Germany, and no one is angry about it any more, but it was a sorry case. Otherwise the Germans are doing their very best to stop the pollution of the Rhine. The Ruhr Gebiet, and indeed their whole industry, is working very hard to keep the river clean. In fact, the French are the worst. All the salts from their potassium mines, the sodium they get as a waste product, they dump into the Rhine. They never go over the official maximum which they are allowed to put in, but they do it continually. We realized what bad effects this had two years ago, when the French industry was on strike and the Rhine immediately cleared up.'

The other major European river whose estuary is in Holland, the Meuse, flows through France, Belgium, and Luxembourg before reaching the North Sea at Holland, and it, too, I learned, is badly affected by industrial pollution – almost as badly as the Rhine – but it is not yet internationally regulated.

Calming down, and returning to Holland's internal environmental problems, Kuenen told me about a relatively unfamiliar pollution source: the innocuous potato, or rather the potato starch industry. 'Up in the north', he said, 'very large potatoes are grown, with a high starch content. They are not eaten as potatoes, but go to factories and are ground up and the starch is removed and used for all kinds of things. You then have a residue, formed of cellulose and some protein, which goes straight into the canals. For over a hundred years, in that area round Groningen, people have known that if the canals smell bad, then the economic situation is good.' Another example of 'where there's muck there's brass', in fact. 'Locally they've got more or less used to it as a symptom of industrial activity. But about eighteen months ago it got so bad that something had to be done about it. So the government decided to build a long pipeline and put the stuff straight into the sea, at the Ems estuary, which forms the border of Holland and Germany. Not surprisingly, this raised an enormous protest, from the Germans, among others, who said: "Look here, it's our estuary as much as yours, you jolly well stop that". The biologists came in very strongly, and did research on the damage it would do to marine life in the estuary, and as a result of their pressure the government has now started to look for a way out. Of course, in government you can't ever simply admit that you made a mistake, you look for excuses; you say, "Maybe we made a slight error in our calculations". In the meantime a lot of money has been given to the potato industry to find some way of using less water, which would make the residue much easier to handle. And they're also looking for ways to use the muck for growing bacteria, yeasts, and so forth, and getting protein rich foodstuffs out of it. Putting it into the surface waters caused eutrophication and general filth and smell, and if they'd put it into the sea, it would simply have killed off all the fish and shellfish for miles. The estuary would have become a dead place, rather like what's happening to Lake Erie.'

Meanwhile, Holland suffers from a more familiar problem. How, in a densely-populated country, with people increasingly mobile and leisured, do you manage your land resources in order to cater for their needs and yet at the same time keep nature reserves undefiled?

'In our Nature Reserves we have all kinds of pollution difficulties', said Kuenen. 'People go in and leave a mess behind; or a neighbouring grassland area is fertilized, and the eutrophicated water runs through the country we are trying to keep in a natural state. It's a continual fight to handle our nature reserves so that they keep their natural characteristics. This is particularly difficult for us because there are three hundred and seventy people per square kilometre in Holland and the amount of refuse is growing all the time. There's another difficulty. We have a very well-thought-out programme of areas we want to buy for national parks and reserves, but the government refuses to put up the money. Because everyone in Holland is supposed to want two cars and two houses and two washing machines, and because raising the money for national parks would involve putting up taxes, the government doesn't dare do it. And so we cannot buy the land we need for research to find out how to manage the rest of the country. We are a bit unhappy about the general political situation in Holland.'

If Kuenen is unhappy about his government dragging its feet in a country which spends more money per head on conservation than any other country in Europe, it is plain how little cause for satisfaction any European can have that enough is being done. But in general, I asked, isn't public opinion firmly behind the conservationists in Holland?

'Yes', he said, 'the public is behind the environment movement, and it's getting stronger and stronger. The big fight at the moment is about roads; some people want to put roads everywhere. There are already two roads between the Hague and Leyden, and they want to put four. When you ask, "What for?" it appears that the reason is to enable Mr So and So, who lives in Leyden and works in The Hague, to start for work five minutes later. That's really all it boils down to. It's just rush hours that are the problem. I've lived in Leyden almost all my life, and I know that if you use the roads between eight and nine in the morning and five and six in the evening they are terrible, but at any other time they're empty. So because of rush-hour traffic they want to cut down woods and ruin large breeding areas of meadow birds. It's really incredible that the road-building people are so narrow-minded and cannot look further ahead than a few years. If you tell them that the people who know about traffic say that the situation will be no better in ten years time, and that all they – the builders – will have succeeded in doing is spoiling the countryside, they say: "Well, that may be true, but we aren't sure, it's just a prognosis, so we're damn well going to put roads everywhere anyway".' Once again, this apparently cool, urbane character practically spluttered with indignation. 'In Holland, we've got good roads all over the place, but they just can't stop making more.

'Then there's another area under attack, the shallow seas behind the North Sea islands off the Dutch coast. Quite a number of people are talking of draining that area for reclamation. This is another example of the absolutely pathological urge of the technologists to go on doing things just for the sake of doing them. Nobody wants it. Agriculture doesn't want it; recreation doesn't want it; the military don't want it; even the water engineers don't want it. But if the conservationists don't clamp down, it will happen and an area of fundamental importance to the bird life of all north-west Europe will have been destroyed. Innumerable migratory birds pass by there and feed; those shallows are highly productive and full of hundreds of thousands of birds – waders, geese, beautiful birds of all kinds. It doesn't freeze over, so they can stay there all winter, or if they are on their way south it's an excellent resting-place. But the technologists, in their quiet subversive way, go on and on. They do a little bit here, and a little bit there and then they say, "We've spent so much money now, we might as well go on".'

Recalling Luna Leopold's preoccupation with putting a recognizable mathematical value on aesthetic considerations, I asked Dr Kuenen for his views. He looked fierce and said at once:

'That kind of work is very important; but the only answer, when people go on asking what nature is worth, is to enquire: "How much is your life, or my life, or anybody's life, worth?" They'll spend any amount of money on saving my life if I'm fool enough to sail out into the North Sea in bad weather and get into trouble. When I signal for help, they send out a helicopter and boats which cost thousands of guilders. There are plenty of times when we don't stop to calculate exactly what something is worth. No one asks what an artificial kidney is worth; medical research is sacrosanct. And that's the kind of attitude we should have more of in conservation.'

9

Emil Hadac

THE ENVIRONMENT UNDER COMMUNISM

AT AN environmental conference in Finland I met a distinguished botanist from Czechoslovakia, Dr Emil Hadac. A small, brown, gnarled man in his late fifties, he had, I discovered, done much internationally recognized work on plant geography; translated Darwin into Czech; lived in Iraq and Cuba as a visiting expert; and undertaken a three thousand mile walk through the Carpathian mountains and Balkans as a young man. In the spring of 1971, I learned, he had been made head of the new and highly original Institute of Landscape Ecology in Prague, which includes not only botanists, foresters, zoologists, and agronomists, but architects, sociologists, lawyers, and economists.

'Eventually, we want to work in many different areas of the country', Dr Hadac told me, 'but first we are going to concentrate on one particular place in north-east Bohemia, about thirty or forty square kilometres, and get all the information possible about the landscape and all the living beings, including the people, within it. We will collect data about the geology, climate, vegetation, and zoology at least of some groups of creatures; a complete survey of insects, for example, would take too much time at first. Then we will get together with the architects and urbanists and sociologists to make a history of human settlements.

'In this part of Bohemia the landscape is clearly differentiated, and you can see at a glance what the differences are. We are taking two neighbouring units and comparing them. In this way we hope to gain some insights into how landscape is made, and thereby to determine which parts of the country are suitable for which uses: forestry, agriculture, industry, or recreation. We must find out how the land can best be used from the economic point of view, for the benefit of man, but not in such a way as to destroy it.'

Hadac is a man of wide general knowledge as well as a first-class botanist, and I asked him why he became a scientist.

'As a boy I became interested, quite accidentally, in archeology. During the First World War, my brother and I – I was five, he was fourteen – were sent out by our mother to find wood and fir-cones to burn for fuel. On our way into the woods my brother said: "You know, there's an old fort round here somewhere, from the time of John Hus, perhaps we'll find something interesting there". And we did find a lot of stuff dating from the early fifteenth century. It was a small castle, which had probably been made of wood. We found several coins dated about 1420, so the dating was clear; the castle was probably destroyed around that time. And I remember that we didn't get home until late in the evening with our rucksacks full of pieces of pottery and so on, but nothing for fuel.' Hadac looked gleeful at the recollection of this childhood adventure. 'And now', he told me proudly, 'our objects are the only ones from that site in the local museum, because later on a lake was made there and it was destroyed. We were just two small boys, our joint age was only nineteen. And that was my start in science.'

Was he brought up in the country?

'No, I was brought up in a small town in eastern Bohemia. After archeology, I got interested in birds when I was at school. My brother by that time was interested in paleontology, but we could not afford to secure literature in both subjects so we decided to find a subject in which we could work together. We chose botany, and bought a book on the flora of Czechoslovakia. My brother later took a job in a bank, but I stayed in botany, trying to discover why plants grow in one place rather than another.'

And where did he first encounter ecology?

'I "discovered" ecology when I went to the university in Prague. I had already been collecting plants, trying to name them and classify them and so on, but with no real system. Then at university I had a number of distinguished professors with ecological ideas and experience of other countries, and I started to be interested in ecological science. I made my dissertation on Iceland, and then in the early part of the Second World War I spent two months quite alone in a tent in Spitsbergen, studying a region of about five hundred square kilometres. It was the best thing I could have done, because if you are living in the same conditions as the plants – my tent was in the tundra, where just below the surface there is ice 300 metres deep – you feel what the plants must feel, how they live, and what they have to endure. I went across the mountains once, in Spitsbergen, where it is 1000 metres high and very, very cold, and I found six flowering plants; one was a yellow poppy, with ice right up to the stem, but quite happy and able to live under such extreme conditions.' Hadac talked of that particular yellow poppy with real affection, and he had also learned a lesson from it. 'When one sees how such plants live, one can quite well understand how they could have endured the Ice Age, in the Arctic, in places not covered by ice. And geologists have shown that such places did exist, in parts of Iceland, Greenland, Scandinavia, Alaska, and Siberia. So we know that plants could have survived at least the last Ice Age in these refuges and we can deduce that new forms or subspecies originated there, and remained in isolation for half a million years. By studying paleogeography in connection with the distribution of plants I could find out very interesting things about the history of plants and their migration. For example, I found several plants, that normally grow only on the sea-shore, in the ice-locked region of Spitsbergen, far from the coast. I wondered how they could have got there, because the plants were sterile; they had no seeds and they were distributed only by streams, and they could not have crept inland against the flow of the streams. Then I looked at a paleogeographical map, and I discovered that in the early post-glacial period the coast was lower than it is now, and the sea reached just as far as that particular area. The plants could easily have got up there, and then, when the sea went back, have been left behind. But they survived.'

Listening to Dr Hadac, one became aware of plants and flowers in a completely new way. From being fragile and ephemeral things, easily trampled and uprooted, they became, through his eyes, infinitely tough and resourceful, ancient and tenacious, and a living, visible clue to old mysteries.

'I find it very interesting, to get together information from different branches of the natural sciences. I have collected plant names, and tried to map them and link them with the distribution of plants themselves. I have found that there is perhaps a connection between very ancient settlements of people, even as far back as the

Bronze Age, and the vernacular names of certain plants. The old names, which have remained the same to this day, correspond with where we know Bronze Age settlements to have been. Later, an amalgamation of newcomers moved in from different parts of Europe, but the old plant names still lived on from generation to generation. In eastern Bohemia, there were two separate Bronze Age cultures, and the boundary between them still exists, in the different names given to exactly the same plants. It sounds fantastic, but I can't explain it in any other way. Such connections, in this case between linguistics and plant distribution, I enjoy very much.

'If you want to understand nature today, you must go to history. If you don't know the historical background you can make mistakes. For example, I studied a blue flowering plant *Hepatica nobilis* or Liverwort. Its seed is distributed by ants, and it does not like high temperatures. When I mapped it in Czechoslovakia I found that it grew practically everywhere on the better soil, in Bohemia and Moravia, but in the Carpathians there were a few small islands of the plants about 45 kilometres in diameter. I wondered why. Then I discovered that in the post-ice age some parts of the country could have been warmer than they are now so the plants could have been destroyed in Slovakia, except for some few places. Then, when the right temperatures were reached again, it could have started to spread again.

'Then there's a river called Váh, with limestone hills on each side, and the right climate for this plant. But it grows only on the left side of the river, not on the right. The only place where it is found on the right bank is on the ruins of a castle, where it was certainly introduced by man, or probably by a woman, cultivating a garden. Why was this? Simply because the ants couldn't swim.' This simple explanation was obviously very pleasing to Dr Hadac and he sat back beaming. 'I have also studied North Atlantic flora, Icelandic flora, and tried to explain where it migrated from in Europe and North America. We had a very fascinating conference in Reykjavik in 1962 on this subject. Up till this last year my work was concentrated on plant communities and phytogeography; I wasn't trying to do anything broader. Now, since I have been asked to lead the new institute, I am going to try to do something more than make a hill of papers. I've already published nearly two hundred papers, so now I can stop and do something sensible, something for society.'

Wherever the conference delegates went during their stay in Finland, Hadac was always among the most active and observant. He was the first to spot a rare orchid in a wood. He was fascinated by the way the Finns manage to incorporate their landscape of pines, birches, rocks, and lakes into their towns and villages, so that there seems hardly any barrier between forest and garden, and was amazed to learn that most forest in Finland is privately owned. He had a strong practical interest in details and a keen concern about environmental danger. I asked him whether there was much public worry in Czechoslovakia about the environment, and whether it was a political issue.

'Certainly, yes', he said. 'I think it's the same everywhere now. Sometimes I think it's been due to the overwriting of certain subjects by some Western authors, who always try to get to extremes, but this was probably the best way to get the subject popularized.'

But did he think the situation was really urgent?

'It is urgent, yes, and if we don't act soon it will be much worse. I'm not as pessimistic as the authors who write that by the year 2000 we won't have enough oxygen to breathe; still, I think we must do something in areas where pollution of the air is high enough to make human life intolerable. And, you see, if we want more oxygen, the reasonable way to do it is to grow plants and trees. But if you want to grow trees, you must plan a hundred years ahead. It is not possible to create a forest in ten years. So we must start now.'

I asked Hadac to identify the most serious and pressing environmental problems in his country.

'Some of the very worst are caused by the paper industry and by textile dyeing. Our air is polluted by power plants. In Bohemia we exploit our deposits of brown coal, which is full of sulphur and also has a certain amount of radio-active material, which can accumulate as sediment in the soil and could be very dangerous for animals and eventually mankind because it could effect the nervous system and genetic make-up. Our new institute is involved in that, too. In two years we must finish a joint project with engineers and chemical experts on a region in north-west Bohemia which is quite destroyed. It has been ruined by coal mining, and electrical plants, and the chemical industry, and pollution is so high that the forests in the surrounding mountains are almost dead; trees live about fifteen years and then die, because the soil and air is so polluted. We are working on this area with engineers and hydrologists and people from a forestry research institute, because it's a very precarious situation and we must solve it very soon.'

But how can problems like these ever be solved while industrial expansion and economic growth remain the over-riding national goals, as they do in communist societies as much as under Western capitalism?

Dr Hadac did not find this question at all daunting. He seemed to think that the problem could be approached in a strictly pragmatic manner. 'We must evaluate the positive and the negative sides of industrial development', he said. 'In our programme, we shall work out what good such an electricity plant or chemical factory does, what it costs and what it provides. We shall include the loss caused, say, to the forests by pollution from this industry; or maybe, the profit, because perhaps sometimes "pollution" makes the fields better: nobody knows. We shall also try to evaluate the effects of pollution on the people, because some of the figures we have, and we don't have very many, indicate that in one heavily-polluted area the population lives about two-and-a-half years less than average. So we can take this fact into account, and include in our calculation the work that could be done in those two-and-a-half years and isn't. Then we can try to combine all this information, and make suggestions.'

So Dr Hadac firmly believed that once the relevant information had been assembled, the government would take the right decision?

'Yes. We must create better possibilities for people. But first we must evaluate what is actually better. I think that in a socialist society you can balance the different arguments more easily than in the capitalist world, where private interests are involved. In our country, industry belongs to the State; the soil belongs to the State, the forests belong to the State, consequently there is not so much conflict.'

But surely, when the State has a strong vested interest in profit and productivity, other considerations are not likely to weigh very heavily with the authorities?

'That is why we must talk to the economic planning people in their own language. We can't just tell them not to do certain things. We must be able to say, "If you do this, then in this region two hundred thousand people won't be able to work properly because they will be sick, and this will cost the country so many millions". Or we might say, "This power plant will produce energy worth so many millions, but will destroy resources worth so much more". Like this, we can persuade them. And if we find that even if one particular area is destroyed, the balance comes out favourably for our society as a whole, then we shall have to say, "All right, go ahead, there's no reason against it".'

Was this why the Russians, for example, were prepared to destroy the Aral Sea, as their spokesman at the conference had indicated?

'Yes. The Russians say, all right, we will lose so many million tons of fish, but we shall get so many fields, which will pay ten times this loss. You have to be realistic. Of course, we must also take into account the aesthetic side, because it has a real value. Why do so many tourists travel to Prague? Because it's beautiful. It is complicated to put a value on aesthetics, but you can work out how many tourists come to Prague each year and how much money that is worth. We can find a language for the aesthetic qualities and I think we will: we must.'

Hadac and his new Institute are trying to persuade the planners in Czechoslovakia to include the environment in their calculations. Were they, in fact, really considering alternative futures for the whole country?

'Well, the inter-disciplinary commission working on northern Bohemia is thinking of the year 2000. By that time, the brown coal will be practically all used up. So what will happen in this district, where there are hundreds of thousands of people employed in coal mines and the power industry? What do you do with a region which is now almost destroyed by pollution? Recultivate it? But how? It will cost an enormous amount. What will these people do? They must do something. Sociologically, it will be very difficult. Our miners are used to a high standard of living, because they get practically as much as a university professor; it won't be easy to persuade them to take other jobs. And there won't be many job opportunities. The region is rich in fuel, but not much else. Yet these people have a right to work; there cannot be unemployed people in a socialist country. So the State must provide opportunities for them to work. In this case, yes, we are thinking very hard about alternative futures.'

This problem, of what to do when resources run out, is a familiar one but I had not heard it stated in such specific terms, or related so directly to individuals, before. It struck me that not only our material but our moral well-being is affected. When so much individual self-respect is involved with work, to allow an old-established industry to work itself to extinction is inhumane, as well as unintelligent. Of course, the problem is not peculiar to northern Bohemia; but, from the sound of it, the Czechs have been taking their situation unusually seriously.

I asked Dr Hadac whether other countries in Eastern Europe were undertaking this kind of work.

'Parts of this work are going on everywhere, among scientists and urbanists and architects' he said, 'but I don't know of another organization which is trying to combine all these disciplines. So from this point of view it's probably the first. Perhaps we are crazy, I don't know. But it must be worthwhile to try.'

10

Two Field Trips

1 THE DISAPPEARING MEADOW

AFTER SPENDING so much time talking about ecology, reading about it, and hearing ecologists describe their various projects, I decided the time had come to take a closer look at what ecologists actually do when they are 'in the field'. Most of them spend much of every summer working out of doors, collecting information which is then studied, analysed, and collated over the winter months.

As I particularly wanted to learn more about the state of the English meadow, which, as I had heard at Monks Wood, was in danger of quietly disappearing, I arranged to spend a day with Derek Wells, a botanist with the Lowland Grassland Research Section, and his young assistant Karen Jefferies, who is married to one of the biologists in the Toxic Chemicals and Wild Life division.

We drove up from Monks Wood into Lincolnshire, where they were to spend two days working in a meadow outside Bratoft, near Burgh. The fact that they had to travel a hundred miles in order to find a suitable meadow tells its own story. Gradually since the war, our old grassland – land that has been traditionally used for grazing and hay-cutting, and that has not come under the plough for hundreds of years, if ever – has been eaten into by intensive farming. Either it has been ploughed up, or it has been treated with chemical fertilisers and sprays, which have drastically changed its character. Because this process has been slow, it is only within the last five years or so that naturalists have begun to worry about the implications. And because there is no immediate drama, no sudden large kills of birds or animals, there is no public agitation. As yet, there is no national 'Save Our Meadows' campaign.

When the old grasslands go, many species of plants and insects go too, and these losses are disturbing biologists and botanists.

Farmers today want good grass as much as ever, but when they use modern chemicals they wipe out many of the less nutritious varieties of grass and plants.

There is another, more subtle result: the colour of the grass changes. As we drove through the thriving Lincolnshire countryside, Derek Wells told me that he could tell at a glance which fields had been treated with fertilisers and chemicals and sown with the new, specially-bred, grasses. 'The green is much darker, and there is more blue in the colour', he said. 'Also, the texture changes; the newer fields are much denser and more uniform in height.' You don't have to be a botanist to feel a profound reaction to the news that the basic colour of the English landscape is being changed in this way.

The Monks Wood team have been searching the country for old meadows in good condition. Where possible, they arrange for them to be bought, either by the Nature Conservancy, or by one of the local Naturalist Trusts. These amateur groups are often extremely effective and the Conservancy likes to work with them as much as possible. If the old meadows cannot be bought, however, there is a new scheme whereby the Nature Conservancy pays compensation to landowners who agree to keep them intact. The Conservancy pays the difference between what the land yields when left to itself, and what it would yield after being ploughed, intensively fertilized, and sprayed.

The mechanics of this idea were laid down in the Countryside Act of 1968, but Wells told me that the Nature Conservancy had only just managed to get control of its first meadow under these terms: 20 acres in Wiltshire, owned by a lady farmer. Earlier deals had fallen through. He explained that it is usually the older farmers who own the desirable meadows, and they tend to be specially tricky to handle, being independent of spirit and not taking kindly to being told what they ought to do with their land, even for the best of motives. He added that the only comparable botanical crisis, in his view, occurred almost exactly a hundred years ago, when sweeping drainage schemes led to many formerly prolific species of marsh plants becoming extremely scarce. At this point, Karen Jefferies remarked how sad she had been recently on coming across a newly sprayed field of cowslips. 'There were hundreds of them, just starting to keel over', she said.

Wells was born in Norfolk – 'the best bird place in the country' – and was a keen naturalist as a child, but gradually his interests moved from birds to butterflies and then to plants. Like a number of his colleagues, from Dr Mellanby downwards, he had started his career unecologically. He had studied agricultural botany, first at Durham University, then at Edinburgh, and spent some years thereafter studying the effects of applying artificial fertilizers. 'I was trying to find out how much grass you can grow by applying nitrogen', he admitted. 'But I salve my conscience by saying that I grew the grass myself that I was working on.' He came to Monks Wood in 1967 to look into the effects of grazing on grassland.

Looking out at the dark green fields we were passing, I remarked on the masses of buttercups. 'The buttercup is quite resistant', Wells told me. 'It can withstand a fair amount of nitrogen, and much of the early spraying didn't really hurt it either. The presence of the buttercup is no real indication of the state of a meadow.'

The Land Rover drew up at the side of a country lane, where the hedgerows were dense with white cowparsley. We got out, unloaded the equipment, which included several large balls of twine, some wooden stakes, and a couple of cameras, and set off across the fields. 'There's our meadow', said Wells proudly. We were met by the local regional officer from the Conservancy's administrative side, who had come to help with the day's work.

The meadow looked particularly lush and appetizing in the early summer sunshine and the green did indeed seem paler and softer, with more yellow in it, than the fields around. The glory of this particular meadow, and part of our reason for being there that day, was immediately apparent. Small purple orchids, *Orchis morio,* with long pale green stems and flat, star-shaped leaves, were in full bloom; and even my inexpert glance detected an unusual number of different small flowering plants among the grasses. 'There will probably be forty or so different plants here', Wells told me, 'as opposed to perhaps six or seven in an ordinary field.'

The main task that day was to count the number of orchids in each of sixty-four plots. Karen was already busy with the twine, marking off long rectangular strips at one end of the meadow. Rows of stakes marked the experimental areas, taking up perhaps one-third of the total; the other two-thirds was being used as a control, for comparison. As well as the numbers of orchids, we were interested in their disposition and the exact position of each plant would also be noted down.

The overall purpose, Wells told me, was to determine the precise effects of different types of fertilizer, natural and artificial, on the composition of a meadow. Meadows with identical soil and situation can show completely different floristic composition, he told me, when differently managed over the years. Like all basic ecological work, this is a slow and painstaking project and he was planning on seven years. 'We are aiming to find a management technique which will enable us to keep this meadow, and others like it, as it is. The old farming practice was to cut it for hay in the late summer, graze it in October and November, and then apply farmyard manure in the winter and spring', he explained. 'In other words, what the farmer took off in hay, he put back in manure, both from the grazing livestock and by hand. Nowadays farmers mostly want to keep their cattle indoors and not let them graze; they can fatten them faster like that, because when they walk about they are using up energy.'

Was he really in search of the best artificial substitute for grazing cows?

'Yes', he said, 'we want to make an organic fertilizer to simulate farm-manure and the manure from grazing animals.' This may sound simple, but in fact, I learned, it involves a complicated series of chemical analyses of soil samples taken from different plots in the meadow, of bundles of the hay cut from different areas, and minute accounting of the different species that grow there afterwards; all these tasks repeated at different times of the year. This kind of work has never been done before. Agricultural researchers have done similar test projects, but their aim has always been to maximize yield in purely agricultural terms; rare plants do not interest them. 'They put these things down as miscellaneous', said Wells, pointing at the orchids. 'It has been known for some time that certain flowering plants, like birds-foot trefoil and clover, contain minerals which are useful to grazing animals, but farmers have found it easier to create substitutes in the form of mineral licks or other additives to cattle feed, and concentrate on growing the protein-rich grasses which will fatten their animals fastest.'

He knelt down to show me some of the other plants and grasses

7. Emil Hadac

8. Barry Commoner
Herb Weitman

9. Paul Ehrlich

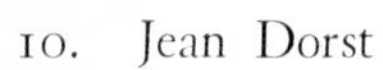

10. Jean Dorst

in the meadow. As well as the orchids, he was making a rougher count of a small, flat fern, known as adder's tongue. It was quite hard to find: 'This fern is particularly associated with old grasslands', he said, parting the grasses carefully with his hands to show me an example. 'It's not exactly rare yet, but it's not as common as it was. The strange thing is that it grows much more profusely on one side of this field than the other.' He pointed out a particularly pretty purplish grass, with a fluffy head, slightly taller than its neighbours. 'That's sweet vernal, one of the delights of this sward, but the farmers don't like it, because it's not too productive. This is rye grass [an ordinary-looking thick green blade] which is very full of protein. This is quaking grass, which you find on chalky soil. These are different types of buttercup: no, I can't find a use for the buttercup.' A small, shiny frog appeared: 'Ah! *Rana temporanea!*' I pointed out some little tree-like plants. 'Those are hawthorns', he said, 'to me they're just weeds in grassland, but my colleague Dr Ward would want to preserve and manage them as scrub. They would over-run this field in a couple of years; but hawthorn bushes are important for birds especially. We all think of something as a weed. It depends on your point of view.'

I asked Wells what he thought should be done if the public felt the need to keep some of the familiar wild flowers that have become weeds to modern farmers, such as poppies and cowslips. It would be hard to justify their preservation on other than aesthetic grounds, no doubt, but I felt a pang at the thought of generations growing up who would know only smooth, homogenised dark green fields. I remembered Rene Dubos's distress about his New York researcher who had never seen the Milky Way. 'I think there's a case for arable weed reserves', said Wells seriously.

We settled down to work in the hot sun. I was given the easy job of simply counting the number of orchids in each plot. Following Karen's advice, I walked slowly up the middle of each plot, looking from side to side. A cuckoo called in the distance, some lambs shouted for their mothers, and a tractor chugged steadily but not too close. There was not an ugly, or indeed a man-made object in sight except for a few strands of wire. We worked steadily, and the day took on a quiet rhythm of its own. I was much slower and more erratic than the others, and several times lost count of my orchids and had to start again. I found it an agreeable, undemanding way of spending a day, and felt sorry for all the people in factories and cities. Gradually I became genuinely absorbed in what I was doing. It actually seemed interesting and significant that one plot should contain a mere seventeen orchids and the one next door eighty-six. I realized that I was enjoying my task so much because the subject was new to me, and because I had never counted flowers before, and above all because the day was beautiful; but I could not help feeling that whereas repetitive and unintellectual work in a factory is stultifying, this outdoor version was soothing and pleasant.

After lunch (a frugal pie and lemonade in a dark, grim pub in Burgh; ecologists, it seemed, were not lavish expense-accounters – Karen and the regional officer had brought sandwiches), I abandoned the orchids and helped Wells find and mark out the half-a-dozen control plots in the rest of the meadow. This turned out to be more of a problem than it should have been. To allow the local farmer who leased the land to graze his cattle without interference from stakes, metal discs had been placed on the ground to mark the corner of the plots. Grass had grown up and covered them, and Wells had brought along a metal detector, like a small hoover, which was supposed to emit a loud whine when passed over the discs, but it turned out that the discs used had been too small to alert this particular machine. 'We used to borrow a mine detector from the army', Wells explained, 'but the local commanding officer left and the new one didn't seem so keen to lend it.' So the regional officer and I spent some time hunting in the grass for the crucial bits of metal. Eventually we found them, and stakes were driven in ready for the next day's work. The afternoon slipped gently away and, unused to so much sun and fresh air, I went to sleep and woke up to find my weight had left a crushed patch of grass and flowers. By this time I was so conscious of the beauty and importance of the meadow that I felt a start of guilt, almost as if I had defaced a painting or torn a page out of a book.

How did Wells go about finding old meadows?

'There are some documents and records kept by local naturalists, and a few old maps and surveys; we usually have some idea from those roughly where to look. But otherwise it's a case of looking on foot, and it's a big job.' The Bratoft meadow had been found like that; and the owner, when they eventually tracked him down, turned out to be a docker in Hull, who had inherited the land and leased it to a local farmer. He had been quite glad to sell it to the County Naturalist Trust. Wells spoke highly of the trusts, who form an important link between the professional ecologists and conservation organizations and the public. Several of the Monks Wood people make a point of sharing their research discoveries with members of the trusts and like to involve them as much as possible in their work.

Before we left, we went across to look at another meadow alongside. Even I could see, although normally I would probably not have noticed, that this meadow was quite different from the one we had been working in all day; it was darker and damper and the plants were quite different. Wells pointed out a plant called pepper saxifrage, which burns the tongue when you taste it, and the regional officer called our attention to a plant called saw wort, which he had noticed on a previous visit and hadn't been certain of finding again. 'I shall sleep happier tonight', he said when he spotted it. There was a profuse growth of a yellow flowering plant, the yellow rattle – usually a sign of poor soil – plenty of meadow sweet, and an attractive bushy plant called Dyer's Greenweed. The professionals, of course, referred to these by their Latin names, but told me their vulgar names after some hesitation.

As we walked, we discussed the work, such as it is, done by earlier ecologists on the flora and fauna of English grasslands. 'Most old-time ecologists were only interested in the chalk grasslands, the downs, in the Chilterns and Sussex', Wells explained. The flowers there were more varied and exciting, and whereas both amateur and professional naturalists were instinctively drawn to the chalk uplands, for bracing walks and family expeditions, the meadows had been left to the farmers, who in any case were not always keen on naturalists rambling about among their cows. A Professor Fream, I learned, wrote a masterly paper in 1880 on the water meadow, and in the mid-thirties a Professor Baker wrote what is still a standard work on Port Meadow, on the outskirts of Oxford. Wells became excited at the mere thought of the riches of Port Meadow. 'It's the oldest thing in Oxford', he said. 'There's a record of 1085 indicating that it was grazed that summer. It's now owned by the City; it has been sprayed, in parts, but there was a proposal not so long ago to spray the whole of it for thistles. That would have been a sacrilege. It would have quite destroyed what had been built up over nine hundred years.' My guilt over crushing the grass where I had been lying was not so sentimental after all, I thought. To an ecologist like Wells, the destruction of an ancient meadow – its slowly-evolved chemistry, and the minute systems of plant and animal life it supports – would be like going to the Bodleian library and burning a medieval manuscript.

2 THE JUNIPER BUSH

The girl wearing the long government-issue mackintosh shivered in the freezing rain slicing across the Wiltshire Downs and gave a juniper bush three sound blows with her stick. As she did so she held out a tray to catch the insects that fell down from the bush. 'That's an earwig. That little beast there is a juniper gall midge. That's a juniper myrid bug, one of the southern species, at a very early stage of development.'

Dr Lena Ward is an entomologist at Monks Wood who has been principally occupied, since 1967, identifying, counting, and classifying, insects that live in scrub land, particularly in juniper bushes. While the Fraser Darlings and Mumfords generalize, in their old age, linking ecological knowledge with the survival of man, Miss Ward counts bugs. What she is doing now, in all weathers, with a kind of subdued fanaticism, is the sort of thing which almost all the best generalizers have done too, when they were younger. Ecology begins outdoors, alone, looking and making precise notes.

Dr Ward, who is a tall woman in her thirties, grew up in a suburb of London. As a child she was fascinated by plants and insects, and her first job was as scientific assistant at the Natural History Museum in South Kensington, helping to sort and classify insects. She did her doctoral thesis on an especially unglamorous and unloved insect called the thrips, a minute black-winged creature which occasionally attacks plants but is not even a serious pest. 'They're a feeble little group', she told me. 'But they've got their points.'

Dr Ward's work for the Lowland Grassland Research Section of Monks Wood is not quite as modest as it sounds. The scrub on which she concentrates is defined, roughly, as all vegetation between the meadow stage and the woodland stage; thus, though sounding prosaic, the scrub habitat is extremely important. All hedges, margins of woods, copses, and young woodlands come under the scrub heading, though until recently scrub has hardly ever been considered important in its own right, because it is either going to grow into a wood, or it is going to be cut back to nothing. But the bushes and tangled plants that make up scrub are crucial for many species of animals, birds, and insects.

Juniper bushes interest Dr Ward because they are becoming rarer and rarer in Britain, and because they support insects which don't live anywhere else. Like many other plants that are now scarce, juniper used to be common and widely distributed, growing especially on chalk downs where constant grazing allowed it the space and light it needs. Now that there is much less grazing on downlands by sheep and cattle, and rabbits are scarce, it has been squeezed out by other scrub plants.

Dr Ward is surveying the location of all juniper plantations in southern England; she also wants to know what kinds of insects live on juniper at various stages of their growth. Like all ecological work, this project is long-term, detailed, and laborious. She notes the age, structure, health, and numbers of the bushes in each colony, as well as their exact site. She takes insect samples from the bushes at regular intervals throughout the year, and at different times of the day. She told me that she sometimes stayed up all night in order to trap a selection of nocturnal species.

I spent an afternoon with her on a deserted hillside near Salisbury. It was cold, windy, and wet. The juniper bushes grew in an irregular thicket along the slope of the downs, a well-established colony, and they were mostly head-height or higher, and dripping. Dr Ward's stick was big and knotted; she also carried a collection of phials and rubber tubes for collecting specimens, a clipboard in a plastic bag, pencils, and a peculiar black canvas tray on a pole, like a flattened-out umbrella.

Each time she found a suitable bush, she thrust the tray beneath its branches, and gave it three blows with the stick. We then crouched down around the tray and inspected the contents. Along with dead leaves, bits of twig, and the berries and needle-like foliage of the juniper, there were what seemed to me to be large numbers of indeterminate small insects. 'Do you mind spiders?' enquired Dr Ward. She assigned the counting and collection of the spiders to me, saying that they were the largest of the insects that interested her and hard to miss. So I started counting spiders, dutifully picking up a representative selection which I dropped into phials of alcohol for later identication.

I was struck by Dr Ward's unflagging enthusiasm. Each time we fell to our knees and peered into the tray she seemed genuinely interested in what we found. A minute bug hopped wildly across the canvas. 'Those are springtailed fleas but there are too many of them to count. That's a young harvestman. They've all got their functions. One doesn't know what all of them are doing, but one knows roughly. This beast here, with the long legs and the reddish body, forms this gall.' Here she parted the needles on a juniper sprig and showed me a minute protuberance that looked like a bud. She opened it with her fingernail and showed me a tiny bug hiding inside.

'Is it a pest?' I enquired.

'Well, it depends on your point of view. If you have too many, it's a pest. That's the thing about nature conservation: it all depends. Insects don't do any harm unless they occur in very high numbers, though anything that feeds on the plant does it a little harm, obviously. This is where the conservation of insects is a difficult topic, unless they're beautiful butterflies or something. But my opinion is that you want to preserve the full fauna on nature conservation "stands" of plants, so that you can see the full range of variability of the animals. On agricultural or horticultural plants you can't take that attitude.

'This beast is one I'm especially interested in because its rate of dispersal is rather slow; it hasn't succeeded in colonizing the younger juniper bushes, though some of them are fifteen years old. They don't seem able to get from here to there.' Many of the bushes we were working on were very old, in juniper terms: eighty or ninety; the younger "stand" was on the other side of a small valley.

'This one is a plant-sucking bug; it only lives on these plants and it's not very common either, only found in southern Britain. This little green blob is a chalcid wasp, which is a parasite. I don't know what it's a parasite of – one doesn't; entomology isn't really very easy. It would probably be a parasite of some caterpillar.' She pointed out an infinitesimal speck of what looked to me like a piece of grey-green lichen. 'This is a bark louse. These feed off the lichens, and as you can see their camouflage is really very good. There are many more of these on the old bushes, of course. That's a fungus beetle.'

She pointed out that some of the bushes were male and some female; the males have small brown cones, the females have light green berries. Some bushes, especially the older ones, have no sex. Dr Ward crushed a few berries in her fingers and sniffed and I did the same, inhaling a powerful and unlikely smell of gin. I imagined Dr Ward having an encouraging whiff when conditions were trying. As we went on round the bushes, beating and counting and noting the results, I became extremely cold; my frozen fingers could hardly manage to pick up wriggling spiders, and the temptation to guess their numbers rather than attempt to count them precisely was increasingly strong: but Dr Ward herself appeared not to notice the cold.

There are no short cuts in ecological research; it always took two-and-a-half hours to do this particular job. Rather to my surprise, although I was uncomfortable, I was not exactly bored. It did occur to me to wonder whether the results ever varied very much and I put this question to Dr Ward. She seemed surprised and told me that only that morning, on the young bushes on the other side of the valley, they had flushed out so many aphids (a common winged insect) that they gave up trying to count them. We were finding hardly any on our side and I suggested we tried the other bushes growing among the juniper. We beat a hawthorn to get a comparison; to my surprise, hardly any insects fell out.

Not far from the junipers, a mobile laboratory in a caravan stood under the trees; inside two assistants of Dr Ward's were labelling insects and juniper cuttings, and filling in the records for the day's work. Another of the Monks Wood entomologists, Dr M. G. Morris, a member of the section, had come down to keep an eye on his work with the small blue butterfly, which breeds on a small flowering plant called ladies' fingers. It was too cold that day to see the small blue in action.

As we packed up and started to walk back to the Land Rover, Dr Ward spotted a large black beetle scuttling through the grass. 'Ah!' she cried, picking it up with great dexterity and cradling it in her palm, 'This fellow can be quite dramatic. Watch.' She started gently teasing the big beetle, tapping its hard domed back with her finger. Suddenly, two bright drops of red liquid fell from the beetle's snout on to her hand. 'That's why he's called the bloody-nosed beetle', said Dr Ward, sounding pleased. 'He does that to frighten you away.' She replaced the insect in the grass and we walked on.

As I settled down in the train taking me back to London, and ordered a large, reviving tea, I pondered respectfully on the long hours, the physical hard work, the repetition and the infinite patience that work like Dr Ward's demands. I thought, too, of the researchers who follow animals around for hours, watching them eat, and sleep, and play, speaking each detail into a tape-recorder. I realized, with a new force, that the backbone of ecology must be the labours of thousands of scientists whose horizons are probably filled with facts and figures about insects and plants. Some of them, no doubt, would move on later to more abstract or more activist concerns. No top ecologist now writing for the general public or serving on a presidential or royal commission on the environment had reached eminence without putting in years of painstaking work like theirs. But those who did not take off into more rarified atmospheres, those who were more than content to go on with their detailed work, would be making the most vital contribution of all to the development of ecological knowledge. The generalizations would mean little without their groundwork and substantiation.

I concluded, too, over the toasted tea-cakes, that the reason why so many people feel drawn to ecology is because it appeals to the thwarted naturalist in them. The grand scheme of ideas is in itself exciting to some; but the real satisfaction of ecology as a science is that it deals in the beauty and intricate detail of nature. Already, amateur naturalists play some part in gathering information; but the work is endless, and there can never be too many people involved, provided that their efforts are properly organized and co-ordinated. With the simplest of briefings, I felt, huge armies of amateurs could be marshalled and employed in the collection of data, which would release the fully-trained scientists to concentrate on the collation and interpretation of results.

But would scientists like Dr Ward and Derek Wells really want to be released? This might remove the reason why they were ecologists at all.

Barry Commoner

THE SCIENTIST AS AGITATOR

DESPITE THE warnings of men like Mumford, Boulding, and Dubos, the fact remains that until recently anxiety over the environment was extremely limited. Hence one could assume that it did not become a matter of public concern as a direct result of the writings of men like these, nor did the conservationist groups begin to attract a wider public until two or three years ago. Everyone I met referred with respect to Rachel Carson's achievement in writing *Silent Spring,* which was a best-seller in 1962, and certainly excited the public – for a while. But how did the environment become a live, political issue? Had the public, as Mumford suggested, reacted spontaneously to their worsening quality of life? Or was there a group of scientists who had been trying to awaken public concern in a more political, activist manner, preparing the ground for the moment when the voters woke up to the fact that pollution affected them as well as birds?

Early in 1970, 'The Environment' made the cover of *Time* magazine, thus achieving popular recognition as a leading trend in American life. The man chosen to represent it was billed as 'Ecologist Barry Commoner', whose bespectacled, earnest face looked out from the cover in an interesting two-tone design. His face was divided down the middle of the nose; one side was black and white against a background of filthy skies, smokestacks, and a single bedraggled bird, while the other was warmly pink against an idyllic sunny landscape. Inside, Professor Commoner had a whole page to himself, headlined 'The Paul Revere of Ecology.' (Paul Revere, it will be remembered, galloped through the night to warn the people of New England that the British were coming in 1775, thus signalling the start of the American Revolution.) If *Time* was right, as it sometimes is, Commoner must be a key man behind the environment bandwagon. I decided to start my enquiries into the activist wing of the movement with him.

Before I went to St Louis, Missouri, where Commoner is head of the Centre for the Biology of Natural Systems, I reminded myself that Commoner is not, and never has been, a professional ecologist; and that he has been regarded by many ecologists with suspicion and perhaps with jealousy.

I flew to St Louis, which lies on the great Mississippi river and calls itself 'The Gateway of the West', on a brilliant autumn day, but came down through a miasma of yellowish brown smog which showed it to be one of the great industrial centres of America. However, Washington University is situated away from the centre of the city, in a pleasant, leafy suburb. The main university buildings had a reassuring, traditional redbrick look, with a clock-tower, wide steps leading up to heavy Gothic arches, and a view of quadrangles beyond. I waited for Professor Commoner, who was delayed by 'some government people', in his office, a small, cluttered room on the ground floor, with a poster pinned to the door showing the Stars and Stripes in bright green, and the slogan ECOLOGY NOW! in large capital letters. This, I learned, had been sent to him by his teenage daughter Lucy.

In his office there were books, magazines, and papers everywhere: on the shelves which covered the walls, on the laden desk, and on the floor. There were numerous filing cabinets and box files; the labels on one cabinet read: Space, Vietnam, Rats, Teach-In, D.N.A., Air Pollution, Fall-out in Food 1960. Among the books I noted *Capital and the American Economy, Poverty and Politics* by Engels, and several new books on the Vietnam war. Commoner came loping in. He is a tall, strong-looking man, with thick greying hair standing straight up from his head like a crew-cut gone to seed. He has black-rimmed glasses, and an expression of great alertness and energy.

'I first learned about the environment from the Atomic Energy Commission', he began with relish. 'That really is the way it was. I was not trained as an ecologist. I'm a biologist, a biochemist, and a biophysicist. Until recently my research work has been to do with the way chemical processes take place in living cells. I don't think I've ever had a course in ecology. But like many other scientists all over the world, after the war I was concerned about the impact of atomic energy and nuclear weapons, and about the consequences of fall-out. It all goes back to about 1952 or 1953 when I became active in the American Association for the Advancement of Science. I wanted to develop a sensible approach to the scientist's responsibilities for the consequences of his actions.' As Commoner recalled those early days his eyes were bright with pleasure, like a seasoned warrior recalling early successes in the field.

'Basically there were two approaches among scientists. On the one hand, there were those who felt that a professional scientific organization like the A.A.A.S. should be concerned only with science, that science presumably has a purely objective content, and that what other people do with scientific knowledge is a social problem, not a scientific one, and none of our business. On the other hand, some of us felt that as scientists we had a special moral responsibility; we were largely responsible for, say, atomic energy, and therefore we should try to influence how it was used. There are difficulties with that approach, too. In the course of our discussions an approach was developed which I had a good deal to do with, which runs along the following lines.

'The decisions that have to be made about fall-out, to stick to that example, are not scientific. Scientific statements can be made about the hazards involved, but the balance between the advantages and disadvantages of developing nuclear weapons is a matter of moral and political judgement. Moral decisions should not be delegated to scientists: they belong to everyone. At the same time these matters have a heavy scientific content, and the ordinary citizen simply doesn't possess the knowledge to know what act of conscience he ought to be performing. Hence it is the scientist's responsibility – and our group came down very strongly on this point – to give the necessary information to the public, so that the public can make the judgements, about fall-out, civil defence, and so on.'

In 1957, St Louis became one of six cities in America where the Public Health Service tested milk supplies for the presence of Strontium 90. In the following year, local scientists and citizens, several hundred in all, formed the Committee for Nuclear Information, which soon became nationally known as the C.N.I. They organized research, collected data from all over the country, held seminars, and answered questions from the public. 'The response was instantaneous, and overwhelming', Commoner has recalled. Soon the C.N.I. found themselves in a position to question the facts and figures put out by the government and they began to have a direct influence on public policy. Most important of all, it became clear that once the ordinary citizen realized that his own food and his own children's health were involved, he stopped being shy of complex scientific argument. A bridge was built between scientists, the public, and the government.

This grassroots activity was clearly a formative experience for Commoner. Projects like the Baby Tooth Survey, when in order to study the effects of Strontium 90 on bone structure C.N.I. had families all over St Louis collecting their children's teeth, proved what could be done.

C.N.I. laid the foundations for what is now the nation-wide Information Movement. Soon after the St Louis group got going, a similar movement started in New York, led by Margaret Mead, the anthropologist, and Professor Rene Dubos. Today, Professor Commoner told me, there are some fifteen such groups all over the country, under the auspices of the Scientists' Institute for Public Information, with headquarters in New York.

Historians of the campaign against nuclear testing give the C.N.I. much credit for the Test Ban Treaty of 1963. By this time, C.N.I. and the groups associated with it had extended their interest to cover other threats to the environment. Perhaps because the nuclear scare is still the ultimate threat to us all, it is hard to take in the fact that it was merely a forerunner of other dangers; and the direct connection between the nuclear fall-out campaign and the general anti-pollution campaign is still not widely understood. C.N.I. itself changed its name to the Campaign for Environmental Information in 1964, which makes the crucial transition clear.

Commoner's account of his own development makes it clearer still. 'My first scientific investigation into the nature of environmental processes was carried out simply by reading the literature about the behaviour of Strontium 90 and calcium', he explained. 'I did it not for any scientific purposes of my own, but to be in a position to educate the public about fall-out.'

How exactly did the transition from fall-out to other environmental threats occur?

'After the Test Ban Treaty, a group of us realized that fall-out was only an example of something much more widespread. The committee of the A.A.S., which I headed, produced a rather important document called the Integrity of Science report. It showed that there were a number of important areas in which science was getting out of hand. In developing that report it became clear to me that fall-out was only one of a number of environmental intrusions which had arisen from the development and thoughtless use of modern technology.'

Naturally enough, Commoner likes to emphasize the role of non-ecologists like himself, Rene Dubos, and Margaret Mead, in the early days of the campaign to wake people up. He can be openly critical of the reluctance of most professional ecologists ten or fifteen years ago to bring their subject out in the open and set about alerting the public to environmental dangers. But this reluctance was not, as Commoner emphasizes, a special failing of ecologists. Almost all scientists have traditionally been nervous of public argument.

Commoner gives full credit to Rachel Carson, who was not technically an ecologist either, for her achievement in writing *Silent Spring;* in fact the A.A.A.S. Integrity of Science report followed up her ideas on the pesticide problem. And he would not be human if he did not quietly enjoy the irony of his appearance on the cover of *Time* to represent the ecologists. But he also recognizes that a small group of professional ecologists, like George Hutchinson at Yale and Lamont Cole at Cornell, were committed early on to furthering the environmental cause.

What, I wondered, should, or could, a thoughtful ecologist have done in the early days?

'Take the insecticides problem. Before the Second World War insects were largely controlled by other insects. What entomologists did was to worry about the life history of insects, and give advice about weather conditions and the cycles that would cause outbreaks of insect pests. An insect ecologist would understand that the steady state population of an insect pest is the result of the rate of reproduction of the pest, the rate of depredation by other insects, and so on. When D.D.T. was introduced during the Second World War the insect ecologists should have said immediately, "If you use this stuff to kill insects you're going to get effects you don't expect, because you're going to kill off predator insects as well".'

During the war, I discovered, Commoner was a naval officer, and took part in experiments with D.D.T. sprays designed to protect American troops landing on Pacific islands from insect-borne disease. 'Our project received all the available scientific reports on D.D.T.', he has said. 'What they told us was only that D.D.T. was a substance with a wholly unprecedented ability to kill insects of all sorts. Some of its other features we learned not from scientific reports but on the job. We learned from a few unpleasant experiences in a jungle in Panama that D.D.T. makes snakes very excited. We learned another lesson when we sprayed an island shoreline in New Jersey and brought millions of flies to that unhappy place – to enjoy the huge mounds of fish that we killed.' Commoner found out not long ago that at the time he was spraying D.D.T. a scientific colleague was discovering the toxic effects of D.D.T. on animals: 'Curiously', he said, 'that information never filtered down to the Navy or to the entomologists with whom we worked.'

That it was left to Commoner and others like him to sound the alarm bells was plainly something he welcomed. He is a political animal as much as a scientist. He has the temperament of a gadfly, and enjoys stinging the herd into action. It is his political conscience, using politics in the broadest possible sense, that motivates him above all. His environmental activism stems from his interest in the welfare of the people affected.

'I was surprised to find that the word ecology had suddenly gotten to be the popular thing', he said. 'I never used it in the early days, really, though I did talk about the environment. If you track back and look at the careers of professional ecologists you will not find many of them going around popularizing the science of ecology because of its social importance, or not until the last few years.'

Was this because the substance of ecology is not easy to popularize?

'But that's what I try to do: popularize. I spend an awful lot of time preaching ecology to laymen. I teach them about ecological cycles; and now I go around telling them my three laws of ecology. I invented these: first, everything is connected to everything else; second, everything has to go somewhere; and third, there's no such thing as a free lunch. That last law is borrowed from economics.

You have to teach people ecology. With the insecticide problem, and later with the nitrate pollution problem, we discovered that the professionals really weren't concerned. They may have known the facts, but the intensity of their concern was minimal.

'About seven or eight years ago I had become so impressed with the degree of ignorance about environmental problems that I raised the issue within the university. I predicted that environmental concern would become very important. At that time I think we had no ecologists on the faculty. I said it was vital to think up some way of mobilizing the scientific competence of the university to do something about the environment and I proposed the organization of what is now the Centre for the Biology of Natural Systems. The university administrators thought I was off my rocker. At that time everyone was saying, the wave of the future is molecular biology, we don't have to worry about organisms in their natural state any more because it's all going to be done with test tubes. Now, I am a molecular biologist; I organized the first training programme in the United States in molecular biology, in that building right there. But while all that was going on, I realized that was not really where it's at. We were learning more and more about smaller and smaller parts of the cell, while terribly important things were happening out there in nature about which we were totally ignorant.'

Could he define more closely the point at which his attitude changed?

'Well, I guess I went through quite a transformation. After the war I really went all gung-ho into the business of interpreting biological processes through physical and chemical mechanisms, assuming that this was where understanding of the complex biological processes lay, that in this little nugget was the secret of life.' As he said 'nugget', Commoner brandished a clenched fist. 'Now, what changed me? It was partly the data; but I had previously developed a philosophical conviction which conflicted with that idea.'

A philosophical conviction?

'Yes, holism. That the real process goes on in the whole complex system.'

I asked Commoner to define what he meant by holism.

'It has a number of origins; the term, of course, was invented by

Jan Smuts, the South African statesman, but it is also closely related to dialectical materialism, if you stop to think about it. The whole idea is to find the decisive relationship in a complex situation, the red thread. It also goes back to the one prominent biologist who has analyzed his method of work, the Frenchman Claude Bernard. Before he died he wrote two books analyzing his own way of thinking; they're extraordinarily informative, and his was essentially a holistic approach. Now the thrust of molecular biology is atomistic, insisting that the explanation of the cell lies in the chemistry of D.N.A., for example. I got involved in that controversy too, and caused great consternation.'

Commoner was one of the few scientists who challenged the Crick and Watson theory that established D.N.A. as the key to the 'secret of life'. He did not do this out of holistic conviction alone; he had years of work on the chemistry of cell reproduction behind him. He accepted the thesis that D.N.A. played an important part in transmitting hereditary characteristics, but was, and is, sceptical about the uniqueness of D.N.A., rejecting the notion that D.N.A. alone determines the features of living cells. In his view, no single molecule is 'the master chemical'.

'Anyway', Commoner went on, 'somewhere around 1960, although I had been doing essentially molecular biological work, I realized that it was going the wrong way so I began marching the other way. But I did not give up or reject my molecular work. I tried to find ways of relating data about the part to the properties of the whole.'

Commoner is proud of his submolecular work, but because of his holistic beliefs he has been determined to relate his researches to environmental matters. 'My submolecular work now', he said, 'is directed towards understanding the whole living organism, rather than just extracts of things. In other words, instead of turning my back on molecular problems, what I try to do is link them up to real biological ones. Just now, the study we're doing of nitrate pollution in Illinois rivers is based on a very esoteric observation I made twenty years ago about isotope fractionation in nitrogen. I'm using a highpowered molecular technique to deal with the fertilizer problem. I see my role now as relating the cutting edge of science to real problems in the real world.'

The Centre for the Biology of Natural Systems was set up at

Washington University in 1966 specifically to research the relationship between man and the environment and to train graduate students. The approach developed by Commoner and his team is essentially inter-disciplinary, or, as he would say, holistic. 'The way we train people around here', he said 'is we all just work together. That's my style: the way I like to work. One of the troubles is that the specialist has a narrow vision. The agronomist doesn't worry about where the fertilizer goes, he only worries about the crop yield. The engineer doesn't worry about where the mercury goes, all he does is to worry about it economically. We have to teach people to worry about the inter-relationships'.

Isn't this what ecology is all about, the study of inter-relationships?

'There is an element of this in ecology but it is really a general description of the way nature works. Unfortunately, the development of modern science has followed a paradigm based on a very narrow segment of nature in which the historic phenomena happened to be very weak: the atom. The atom has been dismembered with huge success. It's parts have been analyzed and the behaviour of the whole atom very well delineated by studying the properties of the separate parts. As a result, atomic physics became the paradigm for all science. Because the atomic scientists obviously had an enormous achievement, everyone else said "Me too!" But as it happens this method does not work very well in all other segments of nature. Take that area of nature that is inside the atom: the nucleus. The nucleus has now been smashed, but no one has been able to predict the properties of the whole nucleus from studying its parts, as was possible with the atom. Some atomic scientists now have a theory, which is really a holistic theory, that the separate parts of the nucleus have properties which don't exist within the nucleus, and do not come into being until they have been broken up – which is exactly the way I feel about molecules and cells and people and society.'

The combination of his holistic approach with the data collected over the last five or six years by the Centre has led Commoner to believe that our environmental situation is extremely dangerous. Although he is not as melodramatic or pessimistic as some of the 'doom-and-gloom' school of environmentalists, he is considerably more alarmist than, for example, his friend Rene Dubos.

'I take it as given', he said, 'that if we were seriously to alter any of the ecosystems on the earth at this moment, it would have very, very serious impact on human life. I don't think we could adapt that well. Let me give you an example. I have talked about the breakdown of the self-purifying systems in water; stress the system too hard and it goes to zero oxygen in the water and the bacteria die, and so on. Now some people would say, all right, there'll be too much organic matter in the water and everything will smell bad, so what? But I ask myself, what happens then? Now two or three years ago I noticed a story somewhere, just a little squib, about a new disease called meningoencephalitis. It turns up in kids when they've been swimming all afternoon, as happens in the South especially. They come home with a headache, they're two days in hospital, and in three days they're dead. It is, as yet, a very rare disease, but it's invariably fatal. It turns out to be due to an invasion of the membranes of the brain by an amoeba, a common soil amoeba. So what's going on? One possibility is that these amoebae grow in the presence of bacteria, but the bacteria don't normally grow in water. But because of all the organic matter now in the water, maybe bacteria can grow there, so possibly the amoebae are stimulated to come out of their cysts by the presence of the bacteria and invade the water. Maybe what has happened is that an amoeba that is pathogenic to human beings, but which normally grows in the soil where it's not very dangerous, is suddenly able to grow in water, a medium which can come into much closer contact with human beings. So ecological isolation breaks down, and you face a wholly new class of disease.

'That's the kind of thing I worry about; and my experience is that when there's a large scale change in the pattern of an ecosystem, then invariably there is a series of serious consequences. It isn't just a question of adapting; one thing triggers another. I think our situation is physically dangerous and I think it's appalling how little we know. And the more we look into the problems, the worse they seem. We don't yet know how badly off we are; for example, the nitrogen thing looked worse and worse the more you got into it, and so did the mercury thing.'

Commoner is quite clear about the origins of our environmental predicament. 'The basic thing that's happened is that a time bomb was planted seventy-five years ago with the scientific revolution.

Between 1900 and 1945 there was a tremendous burst of new knowledge about how nature works; in physics, chemistry, and biology, fundamental new information was collected. Very little of it was applied in practical ways until during and after the Second World War. The classic example again is the atomic bomb, which in 1939 was an esoteric laboratory experiment: by 1945 it was real. In the same way, we knew how to make synthetic organic chemicals in the laboratory in the thirties, but by 1950 we were making them in huge amounts and contributing them to the environment.

'The Second World War was a kind of watershed between the scientific revolution that preceded it and the agricultural and industrial revolution that followed it. Beginning in 1945, we began to do new kinds of things on a massive scale on the surface of the earth. It is absolutely clear – and we said this in the Integrity of Science Report – that in no case did we test what the impact of all this would be in advance. In no case whatsoever. And it looks as though we haven't been lucky enough to get away with it. In fact, by some strange process which I think I'm just beginning to understand, every time we chose new ways of doing things they were invariably worse for the environment than the old ways. It's taken about fifteen or twenty years for the impact to become palpable. It was about fifteen years after detergents were introduced before people started noticing that there was foam all over the place. These things were not degradable, they had to go somewhere; principles of ecology tell us that and the ecologists could have known about it.'

Commoner at this point jumped up and produced a small box of transparencies showing a series of graphs that the Centre has produced, correlating information on various sorts of pollution. He held one up to the light. 'People don't realize', he said, 'that something new has been happening in the world. We've been collecting very interesting data. What this graph shows is a forty-fold increase since 1945 in the use of mercury to make chlorine. Why? Because chlorine is used to make synthetics. What does this mean? Well, for one thing it means we use a lot of electricity to make fabrics which we used to get for free, in terms of energy, from cotton, which grew on water and sunlight. This is a drastic change that just doesn't make sense.'

He pulled out another slide. 'Here's a cute thing: the amount of phosphorus in surface waters between 1940 and 1970. It's gone up from a value of forty to two hundred and sixty. This line represents sales of detergents and this one the entries under water pollution from chemical abstracts and the *New York Times* annual index.' The curves come remarkably close. 'Science should have been able to predict these things from the start.

'Now here's the background data on smog from the Esso engineering corporation.' He produced another slide, showing a table of figures. 'As you see, I have one song: these are just different verses. These figures give the characteristic of the average American engine. Look at what's happened to the compression ratio between 1935 and 1960. It has doubled. I'm responsible for making this point, and I don't think it's ever been made before. I see smog appearing in Los Angeles in 1943, and increasing very fast. My training is to think, "Hell, what's the mechanism behind this?" People say, "Well, we've got all these cars". So I go back and look at the data on cars, and there aren't that many more cars to make that big a change and produce all that smog. So then I look at what's happened to the cars, and I find that the compression ratios have doubled. And then I just sit and think.' Here Commoner gave a quick impression of himself thinking, which meant sitting still and looking unusually stern. 'If the compression has gone up, then the engine is hotter. If the engine is hotter, then the nitrogen and the oxygen in the air and in the cylinders will react faster and produce more nitrogen oxides; and we know that nitrogen oxides are what trigger off smog. And then I go to the engineering literature and I look up the temperatures of how all this happens, and it turns out to be true. Now, is that ecology?' He sat back with an air of triumph.

As a result of his own research and the studies undertaken during the last few years by the Centre, Commoner has reached one basic conclusion about world pollution. The prime causes are the new scientific and technological methods of satisfying human appetites. This is what his graphs and figures are all about. 'We're doing hundreds of these graphs', he told me. 'Take synthetic organic chemicals. Now you see that the per capita production of these chemicals has gone up six-fold, and represents a bigger part of the G.N.P. every year. So I say to myself, all right. What is there about synthetic organic chemicals that makes them pollutants?

Well, for a start, they're synthetic; they are not derived from the natural biological system. There's a whole philosophical thing here. My feeling is that all synthetic combinations have been tried out by living things a million, two million years ago.' My look of amazement seemed to amuse him. 'Yes, some damn fool cell made D.D.T. and found it was no good. But that's a long story; you have to understand biochemistry for that. Mercury exists in the surface of the earth; why don't things have mercury in them? Because the mercury would combine with the sulphur and the proteins and muck things up. The chances are that some of these combinations have been tried already, and rejected: but all this goes back to the origin of life problem. The point I'm making is this; when you make a synthetic organic chemical you are making something that resembles an important element in living systems, but which has been abhorred by living systems. For example, we know that branched carbon chains are not degradable; and the first synthetic detergents we made used branched carbon chains, which led to trouble: there was foam all over the place. And what are the other ecological implications? The social value of making detergents is nil. You can clean things with soap, and the energy for making soap comes free, from the sun. Or take fabric. To make synthetic fabric you have to use up fossil fuel, and it takes much more energy to make a pound of synthetic fibre than it does to make a pound of cotton, counting tractor fuel. What has happened since the Second World War is that we have imposed on ourselves new ways of supplying our needs. The total fabric used per capita in the U.S. since then has hardly changed, but we are wearing different kinds of fabrics.'

This line of thought has led Commoner recently to spend more and more time thinking and talking about economics. The new technologies which do the environmental damage are nearly always highly profitable, he points out, and the hidden costs are borne not by the industrialists but by society as a whole, when they arise in the form of pollution or health hazards. In his latest book, *The Closing Circle,* he tackles the whole daunting question of whether profit and productivity, the twin goals of conventional capitalist and socialist economics respectively, are not fundamentally incompatible with ecological principles. 'No economic system', he says, 'can be regarded as stable if its operation strongly violates the principles of ecology.' In order to put right the technological mistakes which are causing the trouble, he maintains, we need a new set of economic principles.

His conviction that it is above all our technology which is at fault has gradually led Commoner into open conflict with the people who believe that all our environmental problems relate to overpopulation. Recently, Commoner has been making determined efforts to challenge this view, which has led him into open argument with the leading population crusader, Dr Paul Ehrlich of Stanford. In discussing population, it was plain that Commoner had been greatly excited to find that his data bore out a conclusion which he would temperamentally and politically have preferred from the start.

'I avoided the population issue, frankly, until about a year ago', he said. 'I didn't understand it and I had the left-winger's natural uneasiness about it. I used to give talks about the environmental crisis and at the end someone would say, "Why aren't you talking about population?" My answer always was, "I don't understand it, so I'm not talking about it". But a couple of years ago it became clear that this anxiety had merged with the environmental problem, so I sat myself down and I began to study demography. To my horror I discovered that the populationists were totally ignorant, or else unwilling, to talk about what the demographers had learned, which is that population control is not a biological problem. It's a social, political, and economic problem. I began to work very assiduously on it, and I can now show by data that the pollution problem in America is not a consequence of rising population. It is our technology that's wrong.

Ehrlich and I have been having quite a few debates since I began talking about the "demographic transition", as it is called. It is clear that this is the social mechanism for levelling off population. It occurs when the standard of living rises and people no longer want to have so many children. People are concerned about overpopulation for a variety of reasons, ranging all the way from reactionary politics and fear of the coloured races to a feeling that since the ecosystem of which man is part can't grow, the population must be stopped from growing. But overall there is a failure to understand that anything people are involved in partakes of sociology, not biology. That is, it involves politics – which is quite a different realm.'

Not surprisingly, Commoner is one of the few environmental activists who sees the need to put over the message that the environment is not just a convenient consensus issue being exploited by politicians, but potentially a radicalizing influence. In a speech he gave in 1970 in Ann Arbor, Michigan, for the opening of the Earth Day teach-in, he linked the movement with America's two most painful, urgent problems, the Vietnam war and the blacks. He related the war effort to the aerospace and chemical industries, and called it 'The first ecological warfare conducted by America since the attacks on the American Indians'. He went on to say: 'Disaffiliation of blacks from the environmental movement would be particularly unfortunate, because in many ways the blacks are the special victims of pollution . . . blacks need the environmental movement, and the movement needs the blacks.' Commoner is well aware that the people most frequently urged to curb their reproduction rate are the poor and the non-white.

Granted that population growth is not the fundamental ecological problem in America, what about the Third World?

'The world scene, and the problem of the Third World, are something else again', he said. 'What I've learned from the demographic literature about this is very interesting and very shocking, or so it was to me. It goes something like this. Demographers have, I am convinced, described the way human populations regulate their own size with the demographic transition. It happened in England. In the eighteenth century, during the agricultural and industrial revolutions, when enclosures drove people off the land and factories were being built and there were no child labour regulations, it was an economic asset to have a lot of children. So the birth-rate remained high. But because national prosperity increased, and living conditions improved somewhat, and medicine was advancing, the death-rate began to drop. So England went through a population explosion between 1700 and 1900, but after 1900 the birth-rate began to drop and although it hasn't levelled off completely it's pretty close to it. In France and most of Western Europe the same thing happened, and the most recent example is Japan. The demographic transition occurs over and over again; first the death-rate drops, then the birth-rate. I'm convinced that during the rise in population you get a rise in productivity and the wealth of the society increases. It then becomes possible for a working family to support itself if, say, only two people work. Children don't need to work and so you get the introduction of compulsory education. This means that children are no longer an economic asset, but an economic liability. In this country, everyone knows it takes thousands and thousands of dollars to raise a child, which puts an immediate constraint on the number of children people want. If the country has a rising level of expectation, and social security develops, the parents won't feel that they must have an awful lot of children in order to ensure that there'll be someone to take care of them in their old age. For this kind of reason a reduced birth-rate is a consequence of a rising standard of living, and this is what's happened in what we call the developed countries, though I'd rather just call them the rich. They have all, more or less, gone through the demographic transition.'

At no point in our conversation was Commoner happier than when he was discussing his recent thinking on population. It was obvious that to immerse himself in a new area, to find evidence that an established body of thinking was on the wrong track, to gather further data to support this view, and to set out to change established thinking, gave him an enormous kick. Above all, to find his political predilections meshing so perfectly with his new interest was positively intoxicating. 'What about the Third World?' he asked with relish. 'First, the poor countries just happen to be former colonies. Interesting. So you begin to ask, what's the relationship there? Some very interesting studies have been done in Indonesia, and what you discover is that in the poor countries, during the colonial period, the first phase of the demographic transition was set off from the outside. In the West, the new wealth went to raise the standard of living, so that these countries then went through the second stage. But in Indonesia that didn't happen because the new wealth was removed and sent to Holland, where it helped the Dutch go through *their* demographic transition, leaving the Indonesians with no social security, no hope for the future, and with the need to produce more and more children in order to live. As one of the demographers, Keyfitz, put it, the cost of colonialism is one billion extra people. So, in a sense, the population bomb is a natural consequence of the development of advanced industrial technology, and it is a product of exploitation not only of the natural resources of undeveloped countries, but also of their people. Which makes it all the more repulsive and shocking, to me at any rate, that these people are now being told that they must control their population growth, or else.'

Commoner's approach to population is an excellent illustration of his general attitude to the environmental crisis. Ecology, to him, is just one weapon, albeit an important one, to be used in the fight. He is worried by loose talk about ecology and the environment, especially talk that would sweep all problems into that one bag. At the same time, he is beginning to see the need for a new synthesis of scientific knowledge and environmental awareness. Maybe a new role for ecology is emerging.

'It may be time for a new paradigm in science,' he said, 'and the new paradigm may just be ecology, because that certainly is more representative and true to nature as a whole than the atomistic approach which has dominated us for so long. Maybe this is beginning to happen. But unfortunately what ecology is really about is rather confused now, and some ecologists are still really atomists, or what my kids would call "plastic holists", that is, they look at the parts and then say, "Well, we'll just wrap them all together in a computer programme and that will reconstitute the whole". Which is not where it's at. My feeling is that what ecology really means is that we must adopt a sort of natural history approach, study the system as a whole until we begin to sense some key things in it. Then we study those separate elements in relation to the whole, until we see a relationship that carries everything with it. To me, the key to one small problem was that we've increased the compression ratio of our cars. So there is this interesting possibility, that ecology could be the holistic paradigm for a new period of science; that may well be.'

Perhaps, I suggested, people, especially the young, sense this possibility, which is why they have become so excited about ecology in the past few years? Commoner agreed, but added, 'Their enthusiasm also relates to the craze for oriental philosophy, and there are real problems here. One of the easy traps to fall into is to say that nature operates in eternal cycles, and, since everything is connected to everything else, there is really no cause and effect, and no way to intervene without destroying the whole thing. You can fall into a passive admiration of nature, a kind of inactivism, which I think is a mistake, saying that because everything is so complicated, and because you can't understand it, all you can do is admire it. My whole thrust has been to admit and recognize the complexity and do my damnedest to work with the complexity rather than evade it. I have faith that when we see what went wrong, since we made it go wrong we can make it go right.'

How are we going to 'make it go right'?

'There will be no solution, without a major change in our whole productive system.'

But how could people be persuaded to change not only their attitudes, but their material appetities?

'Well', said Commoner with a grin, 'I have a predilection for getting the facts out. I operate on the assumption that if other people understand the facts as I see them, they'll come to the right conclusion. And by that I mean the one that I've come to. What else can you do?' He laughed, aware that he had given himself away. 'But I am actually very attached to the principle of letting everyone make up his own mind. I'm against brainwashing people into thinking that they've got to use bicycles instead of cars, or must get themselves sterilized. The only humane and politically sensible way to proceed is to make sure that the key issues have been delineated, and the relevant facts laid before the public. For the rest, I have faith.'

Paul Ehrlich

THE POPULATION CRUSADER

BY FAR the most startling, and, in terms of public exposure, most effective environmental activist in America (which means anywhere in the world) is Dr Paul Ehrlich, of Stanford University, California. He is a biologist and ecologist who began studying insect populations, became increasingly interested in, and alarmed by, human population trends, started lecturing and writing on the subject and then in 1968 wrote the best-selling book that really launched the population control crusade, *The Population Bomb*.

Since that time, Ehrlich has become a public figure, much more so than any other ecologist. The population control campaign caught on in America like a new religion. The faithful, who ranged from public-spirited suburban housewives to Erhlich's large student audience, received both his writings and his utterances like holy writ; and a national organization, Zero Population Growth, mushroomed all over America, after 1969, to further the cause. Ehrlich himself took to travelling the country in a private plane, and became as hard to reach as an astronaut; he appears frequently on T.V., pounds out articles and books, and contributes to presidential commissions and congressional enquiries. In one year, 1971, he gave over one hundred public lectures and over one hundred radio and T.V. interviews.

The appeal of the Ehrlich approach is easy to understand. First, it appears to reduce the baffling intricacies of most environmental discussions to one simple precept: stop population growth. Second, Ehrlich and the dedicated young assistants he soon collected around him in the Biological Sciences Department at Stanford work at their self-appointed task with immense energy and dedication and a striking absence of the muted, qualifying statements that usually characterize the public utterances of scientists.

Although Ehrlich has won much attention and a following, he has also come in for criticism. Much of this backlash has been the predictable suspicious carping that any scientific community goes in for when one of its number really sticks his neck out; but some of it, it appeared to me, made sense. Although Ehrlich himself is always careful not to blame the poor or the blacks, somehow the population crusade often turns out to be primarily directed at the underprivileged. One does not have to dig very far back in the literature of the population control movement, which after all did not begin with Ehrlich, to discover alarming indications of racial and genetic discrimination; and the social and political implications of what some population control campaigners say now are frequently alarming. Particularly after hearing Commoner on the subject, I approached Ehrlich, accordingly, with some reservations.

Ehrlich came to Denver, Colorado, while I was there, to address a meeting sponsored by the local branch of Z.P.G. The meeting was held in the evening at a huge stadium in downtown Denver, the Old Rodeo Building. Tickets, at $2 each, were available all over town (including the Denver Folklore Centre). The atmosphere beforehand in the bare, brightly-lit hall was reminiscent of a political rally; the podium was draped with Z.P.G. banners and posters, and a large counter was doing a good trade in leaflets, stickers, and buttons: Stop at Two; Suppress your Local Stork; Stop People Pollution; Make Love not Babies. The queues of people crowding in were nearly all young, and tremendously excited by the prospect of hearing Ehrlich speak. I could see very few people over thirty and the average age of the audience looked to be about eighteen, nearly all of them college and high school students, neatly dressed and conventional in appearance. I spoke to several as we waited to take our seats. 'Paul Ehrlich is the greatest man in America', said one boy fervently. 'The population problem is the only real problem in the world today', said another. 'You from England? Well, didn't the Pilgrim Fathers leave England because of overpopulation?' said a third. One youth, who looked about seventeen, told me that he and most of his classmates in high school had already decided to get themselves sterilized, and a group of teenage girls said they had all resolved never to have children of their own, but to adopt babies instead.

On the platform with Ehrlich were a state senator, the president of the local Z.P.G. group, and an aspiring local politician who, after a preamble, 'We have listened too long to lawyers, politicians, and the advertising industry, and not enough to the biologists', introduced Ehrlich as, 'A man who is a prophet in his own time'. Ehrlich stood up amid loud applause.

He is a tall, good-looking, dark man who looks younger than he is; just under forty, he could be in his early thirties. He has a strong jaw and an air of great energy and alertness. He speaks very rapidly, but clearly, with authority, in a powerful, resonant voice. He paid a few compliments to the local politician, said that he had come to give 'a position report on the population – resources – environment crisis' and then launched into a brisk resumé of his established scenario for world disaster if population growth is allowed to continue unchecked. He has a fund of startling figures: 'There are 3·7 billion people in the world, and that's too many; we are wildly overpopulated right now, and the world population growth rate is 2 per cent a year. That means that every three years the equivalent of another United States is added to the world's population.' He made a reference to Vietnam, and to the notion that countries at war need rising populations: 'In all the wars the United States has fought, from the Revolution right through to Cambodia, six hundred thousand soldiers have been killed in combat. World population growth makes that up every three days.'

He hammered home the point that 'population growth is as serious, if not more serious, in developed countries' – or, as Ehrlich sometimes calls them, over-developed countries – 'as in under-developed countries (or never-to-be-developed, in Ehrlich's language). Every American baby born will be a consumer at roughly fifty times the rate of every Indian baby born. Hence, 'there are too many babies being born to middle-class affluent white Americans'. In passing, he threw in a startling statistic to support the view that the people at the bottom of the social anthill suffer most from pollution. 'Our blacks have more D.D.T. in them than our white citizens, because their diet contains more fat and less protein, and D.D.T. is fat soluble.' The message was crystal-clear: 'We have met the enemy and they are us'.

He had the audience gripped and it struck me that there was more than a touch of the revivalist technique in his accusatory style, in the way he brought home to his eager listeners their own guilt and complicity. Self-castigation nearly always has something pleasureable in it.

He moved on to the bald assertion: 'If we don't control ourselves, we will destroy the world', and here he used a neat analogy from his own biological researches. If you observe a population of fruit flies on a banana, you quickly see that there is a pattern: the flies gorge themselves, multiply rapidly, eat out their food supply, and die off in large numbers. This process, he told us, is known to biologists as the 'outbreak-crash' pattern. The human race has had its outbreak; now we seem to be heading for a crash. The crucial difference is that the fruitflies that remain after disaster has struck can always move on to another piece of fruit. But there is only one world. 'As far as we know, there is no other banana.' The audience laughed nervously.

His tone became more deliberately apocalyptic. He effectively dismissed optimistic talk about the extra food to be had from the green revolution (specially bred, highly productive grains) and the much-vaunted 'harvest of the sea'. 'We are jeopardizing the harvest of the sea by pollution, pouring in D.D.T. and other chemicals "which end up killing you sooner or later". So the population crisis is compounded by the environmental crisis: What worries biologists is that we are poisoning the life-support system of the planet. If all the world's food was divided up equally, each individual would get just enough calories, but be undernourished on protein.' He outlined a dismal cyclical effect. Children who have been deprived of protein early in life grow up to be mentally retarded, and so are unable to pull themselves out of ignorance and poverty, and their children repeat the process.

In short, we are asking for trouble and are likely to get it. Another shocking figure: there are more hungry people in the world today than there were people in 1875, less than one hundred years ago. According to Ehrlich, this means that the world's population is weaker than it has ever been. Also the average age is young; there is a higher proportion of young people than ever before; overcrowding is worse; and transportation and communication much more rapid. What this all adds up to is that 'we are ripe for a world-wide virus plague'. So far, we have been lucky. He cited an alarming episode in 1967, when a batch of monkeys sent by air from Africa to a medical research station at Marburg in Germany were found to be infected with a previously unknown virus disease to which humans were susceptible. Thirty people contracted the disease and, in spite of isolation and expert care, seven died. The monkeys had spent a day in transit at London airport. Ehrlich invited us to consider what could have happened if airport personnel or travellers, moving rapidly around the globe, had picked up the virus. There could have been 'almost 100 per cent mortality' from that one incident. Our population crisis, though not responsible for the virus, has created an ideal situation for a catastrophe.

In fact, there is no form of mass death or disaster which is *not* related to the population crisis, in Ehrlich's opinion. He spoke scathingly of the calculations of 'the people at Rand and Herman Kahn', who contemplate with apparent equanimity 'population control by raising the death-rate' and make comparatively optimistic calculations about the recovery potential of the human race in the aftermath of thermo-nuclear war. 'They show a vast ignorance of psychology and ecology.'

So what should be done? First, Ehrlich postulated a tough statement on population control in the U.S. from the President himself, saying that the United States is overpopulated and hence 'patriotic American parents will have only two children. Let's concentrate on quality, not quantity'. At this point, there was a loud burst of applause, and I had a vision of Ehrlich becoming President on a Population Control Platform, swept into power by teenage voters. But fewer children alone would not be enough. 'Our society needs to turn around. There are a lot of things we need in this country other than big flashy cars'. He paid tribute to Kenneth Boulding, at Colorado's leading university, and his ideas for a spaceship earth economy.

Finally, he came back to the greedy, selfish behaviour of rich countries like the U.S. 'Six per cent of the world's people using forty per cent of the world's resources.' It couldn't go on. 'What we are doing', he said, 'is stealing from people and holding a gun at their heads while we do it.' Why is the U.S. in Vietnam? Because she wants the resources of South East Asia. 'We're investing ten million dollars in petroleum exploration in South East Asia in the next ten years.' Look at the flashpoints in the world situation today, and you find vital resources are nearly always involved. Look at the Middle East. Over-population means greed which means international tension which means war.

11. David Gates

12. Kenneth Watt

13. Edward Wilson (right) with Richard Levins

He ended on a somewhat gentler note. We must stop behaving badly and start behaving well: 'The only realistic solutions today are idealistic solutions.' Prolonged applause. 'Americans must stop thinking about politics as a dirty game. It's up to you. I wish us all luck. Thank you.'

I came away from the Rodeo Hall both impressed and unnerved by Ehrlich's brilliant performance. Was it really necessary, or wise, to be quite so apocalyptic? Was there not a likelihood of such sweeping alarmism being counter-productive? Although his concluding message – that America must re-think its exploitative attitude to her own and the world's resources – seemed wholly admirable, there were statements and assertions in his speech which could with reason be attacked as misleading, inaccurate, and socially unwise. To say that D.D.T. kills people sooner or later is certainly a wild exaggeration as far as present knowledge goes. The imminence of raging virus plagues decimating the globe seemed overstated. More seriously, I could not rid myself of the uneasy feeling that whereas in context his remarks about protein deficiency impairing brain development sounded innocent enough, taken out of context they could be construed as an assertion that poor, ill-fed people are stupider, and therefore less use and more expendable, than rich, well-fed people. Ehrlich's speech was great propaganda, but I wanted to know more about him personally, and about the development of his ideas.

One thing is clear: Ehrlich has an absolutely solid scientific background. He has been Professor of Biology at Stanford since 1966. He was born and brought up in Philadelphia. His father was a shirt salesman and his mother a schoolteacher, and they both encouraged his early interest in nature and insects. He studied biology at the University of Pennsylvania, read William Vogt's book *The Road to Survival* in 1949, just after it came out, which was 'a major influence', and specialized in entomology and the study of insect populations as a graduate student at the University of Kansas. One project he was involved with was a study of the genetics of D.D.T. resistance in fruit flies. He was also fascinated by, and worked on, butterflies, which, he has said, led his father to think 'my professional career was basically going to be silly'. His father died while he was in graduate school. He met his future wife, Anne, at Kansas and she became a biological illustrator, and is now one of his research assistants in Biology at Stanford, and works closely with him. They produced a book on butterflies together in 1961, *How to Know the Butterflies,* and she is co-author of his recent book, *Population, Resources, Environment,* a lively, exceedingly thorough textbook which collates all the latest information on world environmental problems and weaves it all into the Ehrlich message. They have one daughter, Lisa, now in her teens. Ehrlich has himself been sterilized.

So his professional interest over twenty years has always been, as he put it, 'The theory of population biology and the genetics of population dynamics'.

How did he become a campaigner for population control?

'I just got more and more concerned with the inter-relatedness of things, and started worrying about how we were going to feed mankind without destroying the planet. I've always had a strong interest in man as well as insects. I did some work on the ecological problems of the Eskimoes, which got me interested in the human side.' He went to Stanford as an assistant professor in Biology in 1959, and 'started teaching an evolution course, which got me thinking about where man comes from and where he is going, and the implications of the agricultural revolution. Then in the early sixties I started talking to student groups about our problems, and how they were going to finish us off unless we did something about it.'

He traces the beginning of his career as the number one population crusader to a speech he made to the Commonwealth Club in San Francisco in 1967. 'They had radio and T.V. coverage and I got a lot of attention. I started talking regularly on population problems, and David Brower of the Sierra Club heard me and asked me to write a short book for them. And my life was destroyed from then on.'

Ehrlich himself, I discovered, has mixed feelings about his subsequent fame and public exposure. When I talked to him, he sounded frantic. He had a week of speaking around the country ahead of him, then he was going to Washington to testify on population problems before a presidential commission, then he was off for five weeks to Panama and Trinidad to do some field work. Speaking of the handful of scientists like himself who lead the environmental activist movement, he said: 'In recent years the notoriety and the seriousness of the problems has led to enormous pressures on us. On my field trips I do a very different kind of hard work. That's the kind of work I really love.' He has worked in an impressive range of places: the dustjacket of a recent book says he has 'conducted field research in Colorado, Alaska, Mexico, the Canadian Arctic and Subartic, Australia, New Guinea, New Britain, the Solomon Islands, Malaya, Cambodia, India, Kashmir, and East Africa'.

What kind of work does he pursue on field trips?

'I'm still studying insect attacks on plants and plant response. We'd like to get the information necessary to design reasonable agricultural systems in the tropics.'

It was his trip to South East Asia and India on a sabbatical year in 1965–66, which, he has said, 'solidified my view of the world population situation emotionally. The drama of what's happening in places like India is really overwhelming. In the United States you can confront poverty very easily near almost any large city or out in the Navajo reservations, but you can also leave it very rapidly. In India you seem to be surrounded by it and it's something you can't ignore. It's easy for Americans to ignore poverty, because the superhighways don't go through the shantytowns; but when you travel in the rest of the world, the desperate poverty of the people faces you everywhere and it gives you an emotional feeling for the humanity trapped in these problems that you can never get from U.N. statistics.'

In talking to Ehrlich and reading his work in more detail, I realized that he is calmer and more moderate by several degrees than his speeches, or a deliberate piece of propaganda like *The Population Bomb,* might lead one to suppose. He acknowledges, for example, that problems like the arms race and poverty in cities and racial tension have to be tackled in their own right, not simply as an extension of the population crisis. Also, recent criticisms from men like Barry Commoner seem to have caused him to modify his views, although he openly disagrees with Commoner's contention that in the U.S. at any rate it is technology rather than too many people that is the main environmental problem. 'He is dead wrong and I can prove it. Increased population growth increases per capita power use; you can't get away from that.' But he now always emphasizes that population control must be combined with a re-adjustment of technology and an all-out attack on pollution. He told me: 'I agree with Commoner when he worries about the political implications of what I'm saying. I worry about them myself. We all have to worry about them.'

There is no doubt that, in his public pronouncements, Ehrlich has made a deliberate decision to sacrifice the detachment and measured analytic caution traditionally expected of a scientist, in order to rouse people's emotions and energies as his own have been aroused. He and his associates believe in shock tactics. They do, often, overstate in order to make an effect. One stunning example is to refer to pregnancy as 'a nine-month disease'. To my mind, population control, with its intimate human implications, needs more delicate handling. But it is hard to blame Ehrlich. He and his disciples have been driven to overstate, to go around alarming and upsetting people, precisely because the traditional approach, as employed by scientists and politicians alike, has proved inadequate. The fact that population control needs to be handled with care has tended to mean that it is not handled at all. If even relatively secure countries like the U.S. and Britain, with excellent communications, have no firm population control policies, what hope is there for the rest of the world? A vicious circle operates; blindness, frequently combined with religious prejudice, and the nervousness of scientific and political leaders, means that little concrete is done about population growth. So people like Ehrlich are forced to overstate a case which really needs no overstatement, as he would be the first to admit. Their efforts create a stir and stimulate population concern, but also create an inevitable backlash among scientists and politicians, so that still nothing is done. Both Ehrlich and Commoner realize that it would be highly unfortunate if discussion of the population problem were to be overlaid by argument between them, but by early 1972 their difference of opinion had become open conflict verging on hostility. However, Commoner knows the population crisis is real, and Ehrlich knows that technology must be controlled. An important point, it seemed to me, having talked to both of these men, is that the population problem should be brought now into regular politics and made part of the regular political diet of the overdeveloped countries, and not left to burgeon on its own.

13

Norman Moore

A BIOLOGIST IN WHITEHALL

THERE IS no equivalent in Britain or Europe to the campaigning ecologists like Barry Commoner or Paul Ehrlich in America. This is partly because the environmental crisis is further advanced, and more obvious, in parts of America; partly because of differences in scale; partly because of different national temperaments; and partly because scientists in Britain and Europe are by tradition and training a more academic, inward-looking group, who on the whole greatly prefer their laboratories and professional colloquies to public platforms.

However, this does not mean that no progress is being made in drawing attention to and tackling environmental problems. In some respects the British and Europeans, in a quiet and undramatic way, are ahead of the Americans on such practical matters as the banning of certain pesticides, or air pollution controls. The activists to be found among the professional ecologists in Britain and Europe are seldom as startling or inspiring as their American colleagues, but they manage to get an impressive amount done.

As head of the Toxic Chemicals and Wild Life team at Monks Wood, Dr Norman Moore has probably done more than anyone else in Britain to alert professionals and the authorities alike to the dangers inherent in the indiscriminate use of pesticides. Although he is not really a popularizer, he has hammered the message home for ten years or more – where it really counts, among his colleagues and peers in science and agriculture – in a succession of learned and detailed papers and articles, participating in and organizing innumerable conferences, and playing a leading part around Whitehall in advisory committees with government and industry.

Moore is a tall, concave man, with long, thin, legs and arms, a gentle, ascetic face and nervous darting movements. His style and manner resemble that of the traditional university don and it comes as a surprise to hear him discussing matters of urgent topical importance rather than a point of scholarship. Unlike his colleague Mellanby, he came to conservation comparatively early in his career and has been actively involved with the Nature Conservancy for nearly twenty years.

It did not take him long to discover what his life's work was to be. 'I was brought up in the country', he said, 'in Sussex and Kent, and all my life I've got enormous pleasure out of looking at plants and animals. I would reckon this was one of my main motivations. I am just extremely interested in wild animals in their habitat. I was a fanatical bird watcher as a boy. I was also a fairly fanatical hunter. I used to do a lot of wildfowling.'

Moore read zoology at Cambridge, and after the war went to the University of Bristol as a Lecturer. He had no formal training in ecology, but pursued the study of the territorial behaviour of dragonflies, a subject on which he is a world expert.

But how did dragonflies connect with conservation and ecology?

'I had always been concerned with how behaviour affected the ecology of a creature. After a bit I got more interested in the ecology and less in the behavioural side. Gradually I got particularly interested in the effects of human beings on natural systems. I felt it was quite absurd to think of the natural world as a backcloth against which human beings were acting. To me, the interesting thing about ecology was that man was part of the system. But I found that in universities as soon as you began talking about applied problems people became uninterested. The academic attitude was: this is applied science so it is second-rate by definition. It seemed to me, by the early fifties, that the interaction of man and the environment was absolutely crucial and that we ought to be studying it, but because of a sort of intellectual snobbery nothing was being done.

'I was always interested in conservation in the narrow sense, how to preserve wild life, and when an opportunity turned up I joined the Conservancy in 1953.' He became their first Regional Officer in south-west England, based at Wareham in Dorset. This meant he had to select and look after Nature Reserves, organize and co-ordinate research, advise planning officers. 'It was a scientific administration job, but over a wide range.'

Moore was undeterred by academic resistance to his ideas. He spent seven years in Dorset, and produced a long paper at the end of his stay called *The Heaths of Dorset and their Conservation,* which contains a detailed record of years of patient field work as well as general observations on conservation principles. 'I thought the whole question of man in the ecosystem was worth following up', he said modestly. 'I think this was one of the first studies in this country relating human activity to the scientific aspects of conservation. It describes what happens when you reduce the size of heathland areas.'

He showed me two maps. 'This one shows the extent of heathland in the nineteenth century when Hardy was writing his novels.' A substantial part of the map was shaded in. 'This is what it was like in 1960.' The heathland areas were much smaller and broken up into small separate blocks: farms, towns and timber plantations had eaten into it. 'I was studying the effects of this fragmentation on the fauna, showing how species in these small isolated parts die out, and why. Though the original type of vegetation hasn't changed, there's so little of it now and it is so fragmented that accidental extinction of species occurs in outlying places, and this is very fundamental: nature reserves must be large enough to protect viable populations of their flora and fauna.' Moore chose ten indicator species – two plants, two dragonflies, two butterflies, two lizards, and two birds – and studied them in different spots over five years or more. His investigations also covered, among much else, the disappearance of heather, the increasing scarcity of the Dartford warbler, the survival of some species in new habitat and the reappearance of deer.

It was the method and overall approach of his study that distinguished it from most field surveys. He was concerned with a whole habitat, not just bits of it that happened to fall in the Nature Reserves under his control, and he was determined to combine biological detail with searching analysis of what conservation should be about. 'Conservation', he wrote, 'contrary to what is often supposed, involves research, prediction, and control, rather than laissez-faire. The definition of aims is an essential forerunner of these activities. . . . The original habitat is disappearing rapidly; it will lose much, if not most, of its interest unless conservation measures are taken. Is its scientific value great enough to merit them?'

He also made it plain that the fate of the heathlands, and their flora and fauna, was inextricably tied up with the pattern of human activity in the area. But his main aim was, by studying one habitat in painstaking detail, 'to obtain a scientific basis for administrative action'.

After this thorough grounding, Moore moved on to study pesticides. Although he did not work full time on pesticides till 1960, he had dealings with them before that. In the early fifties, in south-eastern England in particular, numbers of birds died, affected by DNOC, and by the highly toxic organophosphorus insecticides that had been sprayed on cabbages to control aphids. Organophosphorus compounds include Parathion and Tepp and are extremely dangerous to mammals, birds, and man; a single ounce of Tepp could fatally poison nearly five hundred men. But they had one advantage: they were not persistent, and so their effects were limited. Moore did some field work for the group formed in 1954 by Sir Solly Zuckermann, the government's scientific adviser, to evaluate the effects of some of these chemicals, which led to their being much more strictly controlled. Then in the late fifties and early sixties, more large kills of birds were reported, especially in the spring, after sowings with dressed cereal seeds. This time the culprits were the organochlorine group, dieldrin, aldrin, and heptachlor. At this point the Toxic Chemicals and Wild Life division of the Nature Conservancy was formed, and Moore was asked to head it.

'By 1959 we were beginning to get severe casualties of birds, and the Conservancy thought it was high time we had a unit that could really find out the facts about pesticides and advise the government what to do. At that stage practically nothing was known about ordinary wild life in farm land, where pesticides were obviously having most effect, so it was then that we started our work on hedges, for example. No one knew anything about hedges at all. We started a study of birds and plants in hedgerows. At first we looked at them from the angle of spray drifting on to them, but very soon we were looking at them in their own right. We found that hedges were disappearing, which raised the question: does this matter from the agricultural point of view? And then: does it matter from the conservationist point of view? That was how the unit gradually became concerned not only with the effects of pesticides but with all agricultural practices. Later we became concerned to some extent with other pollutants as well – those resulting from industrial uses of chemicals.'

The Monks Wood team was one of the first ecological interdisciplinary teams ever to be set up, and Moore takes considerable pride in his pioneering. 'It was very exciting. I had always been frightfully keen on running a mixed discipline department. I used to dream about it in my university days, but it is quite impossible to do it in universities in this country because of the teaching structure. You've got to have a chemistry department, a biology department, and so on. But here we were able to look at it from the other end. We could say – we want to find the answer to this general problem and for that we need ecologists, toxicologists, entomologists, and so on. In 1960 this was really a new concept. This was long before Rachel Carson wrote *Silent Spring*. People often think we started because of that book, but they telescope history; by the time *Silent Spring* was written and read, we'd already got going.'

Rachel Carson's famous book which came out in America in 1962, and in Britain in 1963, certainly did more than anything else to wake up the general public in America and Europe to the hazards of pesticides, and always comes up in any discussion of the subject. Although all the professionals give the book its due as a propaganda landmark, I thought I detected, among the British scientists, a certain reserve. For one thing, they were somewhat alarmed that Miss Carson had scarcely a good word to say for pesticides at all. 'An advocate's case; she only put one side of the question, though, of course, she did it brilliantly' – this was the sort of comment I heard again and again. The fact that the book has become a kind of sacred text obviously offended the judicious, cautious professionals like Mellanby and Moore, who feel it is most important to keep emotionalism at bay.

Moore looks back in astonishment that so much is now taken for granted which ten years ago was fiercely contested. 'In the early sixties, up to about 1965, I had the most tremendous battles. I was quite determined to get over one idea in particular, the question of persistence. I spoke at the British Association, and it caused quite a stir that we were taking this critical line. Immediately, we ran up against big business, and also what one might call the agricultural establishment, people who didn't like the idea that something as useful as D.D.T. could have any snags at all. They simply didn't want to believe it; a lot of them had staked their reputations on saying D.D.T. was really quite safe. It was a very exciting time, and we really pushed quite hard. We lectured like mad and took every opportunity to influence people and, of course, in recent years our efforts have borne fruit very well. It's almost unbelievable that nowadays these problems are recognized by thousands of people, whereas ten years ago only a handful of us had grasped concepts such as the significance of persistence at all. When we started work everyone was interested in toxicity. The chemicals that were generally recognized as harmful were the very poisonous ones like Parathion and the other organophosphorus chemicals. But I felt quite sure that the stuff that remained in the environment must matter most of all. I remember around 1963 deliberately introducing the phrase 'environmental contamination by pesticides' and people would say, what on earth do you mean? Pesticides aren't *pollutants*. And yet already we'd had one or two pointers. I was very struck by Hunt and Bischoff's work on Clear Lake, though the account was hidden in an obscure Californian journal in those days.'

The Clear Lake episode is the classic example of how persistent insecticides work. In 1949, D.D.D. (a chemical related to D.D.T. but rather less toxic to fish) was used in the lake to deal with a tiresome gnat problem. The aim was to kill off gnats' larvae, which live in water. At first the operation seemed extremely successful; the gnats almost disappeared and there appeared to be no damage to fish and wild life. After 1950, the gnats returned, gradually creeping up to their previous level, and further applications of D.D.D. were made. But in 1954 large numbers of dead grebes were found at the lake and after some public uproar and several years' work, the facts emerged. It appeared that even the first application of D.D.D. had affected the reproduction of the grebes; but the deaths were thought to be caused by the D.D.D. travelling up a food chain, from the plankton and other minute organisms in the lake, to fish, then to the grebes which fed on the fish. Each successive form of life that consumed the chemical concentrated it, so that eventually the grebes took in a killing dose. This case was a landmark in the pesticides controversy, although it is now known that fish usually obtain most of their pesticide residues from the water through their gills.

'I remember thinking, well, now we've got what looks like accumulation of D.D.D. in an aquatic system. Then there were some analyses which made it look as if birds of prey might be getting rather more of the stuff than other birds. So we began looking into it, and this originated the whole study of pesticides and birds of prey. We did all the earliest work on this and followed it up, and it has been a very important story – one of the very few cases in Britain where there's no doubt whatsoever that a pesticide has damaged something through at least two stages in a food chain. It's usually very difficult to trace these things.'

Moore's team started investigating birds of prey in 1962. The populations of certain predatory birds had declined dramatically since the late fifties and the birds' breeding-levels had been falling since the late forties. They studied, among others, the peregrine, the sparrowhawk, the barn owl, the golden eagle and the heron, and patiently built up a cast-iron case against persistent pesticides. As with the grebes in California, it was via their food that the birds of prey were affected. They collected disturbing evidence to show that, besides killing birds, residues of D.D.T. or allied substances had strange effects on the birds' reproductive mechanisms. The calcium content of eggshells was reduced, so that the birds had great difficulty in incubating their eggs without breaking them. Certain species, notably the golden eagle, were consistently laying sterile eggs.

Although much work has been done elsewhere on pesticide effects on predatory birds, particularly by university scientists in Wisconsin (Aldo Leopold's old department), it was pioneered by Moore and his group. Moore is anxious to give his colleagues full credit. 'Ratcliffe was the first to get on to the eggshell business. Jefferies has now got fairly near to discovering how D.D.T. affects reproduction, through the thyroid. Monks Wood really has been very much the centre of research. Many of the major developments started here.'

It is one of the unique features of the British set-up that a man like Moore, whose approach is essentially critical, can also be involved in putting things right. Moore obviously thinks this is very important. 'Ever since it became clear that pesticide use must be better controlled, we've had the Pesticides Safety Precautions Scheme, a voluntary arrangement between government departments and industry to decide which chemicals can be used and how. Since 1961 this has been the channel through which Monks Wood formally gives its scientific advice. I personally have been very much involved with improving the routine system of pesticide control in the country. Once a month I go to Whitehall and meet people from the Medical Research Council and the Agricultural Research Council, and Government Departments and we look at information submitted by industry about new chemicals before they come on the market. We all sit round a table and make up our minds about them. So as well as bellyaching about the system, we've done all we can to improve it.'

Like Mellanby, Moore is unnerved by some of the passion and acrimony coming from America about pesticides. 'In Britain, we don't believe that all industry is automatically wicked; we say there are some very wicked men in industry, but there are also people who really do want to do the right thing, and we try and tie up with them. We learn to identify these people, and they can do an enormous amount of good. Our voluntary scheme has, in fact, worked well already. Largely owing to recommendations from Monks Wood, based on our research with birds of prey, and on the pollution of fresh water and the sea by pesticides, many uses of aldrin and dieldrin were stopped voluntarily, and there has been a marked improvement all round since 1962.'

But, as always with ecology, there are no short cuts or instant diagnoses, let alone instant solutions. 'The sudden popularity of ecology is horrifying sometimes', Moore told me. 'People wish there were short cuts and there aren't.'

It had struck me when the ecology boom broke that what excited people was the handful of general statements which, though apparently simple, in fact represented a great deal of long, hard work. People absorbed these lessons and now they wanted more, not quite realizing what was involved.

'That's very true', said Moore. 'People seem to expect a new ecological story every week; but ecology isn't like that. It is very unlike chemistry, where one experiment can produce a clear cut conclusion. Ecology is complex and long-term. But at least we have finally managed to get many people to look at life in a rather different way, a more realistic way. The trouble is that all these pollution problems are so urgent. It's a frightful battle against time, and we have to make decisions again and again on inadequate evidence. It's all very well for people in ivory towers to say, "We haven't absolute proof of this or that", but doing nothing can be so disastrous that one is *forced* to act on inadequate evidence.'

Obviously Moore has suffered from this particular criticism. He became almost agonized as he described the dilemma of the public-spirited scientist. 'A lot of conservation biologists have made themselves very unpopular. They know jolly well they've made a decision on bad evidence, but they think it's better to make a decision like that than to do nothing. This is where some scientists *have* to be politicians. They are in a cleft stick, their whole professional training is against making decisions until they're pretty certain; and yet, if *we'd* waited until we'd dotted every pollution "i" and crossed every pollution "t", the North Sea would be sterile, and Britain would have no birds of prey. Laissez faire is a hopeless approach.'

So he had found this dual role a problem himself?

'Very much so. Anyone who feels the need to see that his research is applied, who is not just a pure researcher, is in this predicament. And we're very vulnerable to criticisms from our scientific colleagues who are happy to sit back and say: "Oh well, it's up to the politicians".'

Had his need to take a decision ever led him astray, or put him in a false position?

Moore looked very thoughtful. 'I don't think it has quite done that. But frequently one has to pronounce on things one would rather defer judgement about. Whereupon there are always people around who say one has been unscientific.'

Thinking of Commoner and Ehrlich in America, I asked Moore why some scientists chose to plunge into the public view, and others did not. Was it simply a matter of temperament?

'All I can say is that I'm very much aware that *my* primary motive is conservation. I want to get something done about conservation and so in a sense my research is a means to an end. When I'm studying dragonflies, all I want to know is what really happens, but even here I want to apply the results if they help conserve dragonflies. A lot of other scientists are really only interested in understanding the natural system. They could hardly care less what was done as a result of their research. Both sorts are needed.'

14

Jean Dorst

SAVING THE MARSHLANDS

THERE are few ecologists on the continent of Europe, and there are still fewer who are interested in fighting conservation battles and waking up the general public. Jean Dorst, Professor of Zoology at the National Museum of Natural History in Paris, is one of them. He is a thin, dark man in his fifties, with a faintly ascetic air. Unlike Kuenen, he seemed a private rather than a public man. He spoke of his own work with passion; the study of ecology was obviously an intense personal experience for him. I felt that at any other time in history he would have been by choice a pure researcher, content to leave the organizing and campaigning to others but that today, in France, he was urgently needed. And so he is a tireless campaigner in France and through the I.U.C.N. He is also President of the Charles Darwin Foundation for the Galapagos Islands, an organization that he helped to start. In addition, he is Vice-Chairman of the International Council for Bird Preservation, and Secretary-General of the Zoological Society of France. And occasionally he gives ecological tutorials to President Pompidou.

Like Mumford and Boulding, Dorst sees ecology as inextricably bound up with the study of evolution. 'I think we, as biologists, have to face one important problem, and that is evolution. And to me, the most interesting approach to it is via ecology. We need to see not only how one species evolved through geological time, but how all species developed side by side, in competition with each other, influenced by predation; in fact, how an entire ecosystem built up. That is probably why I became an ecologist.

'If you look at the history of biology, you will clearly see that it was a science that began by analysis. This is still the common approach. During the last century, biologists tried to make an inventory of all living species. They had most success with birds.

The systematics of birds is much more advanced than that of any other group of animals. This was the first stage. Then came the second stage, the study of evolution from the point of view of the anatomist and the morphologist. So scientists began to build up the phylogeny of groups, according to their anatomical and morphological characteristics. And the biologists succeeded brilliantly. I need only refer to what the British school of evolutionists did. Of course, I am a Darwinian.'

So Dorst looked to Darwin as his intellectual hero, rather than the great French biologists Buffon and Lamarck?

Dorst replied that, of course, both these biologists were great men, but they were part of the first, descriptive phase of evolutionary studies when compared to Darwin. 'Charles Darwin was a great scientist and he greatly influenced my own way of thinking. Of course, his ideas are already a hundred years old, and he did not know all the facts that we know, so he had to make some mistakes; but he had an insight which was unsurpassed. I would say that the great names of evolution, like Darwin and Huxley, were already in a way ecologists. It was not called ecology, but they put evolution in the context of natural elements. Ecology certainly didn't start in this century, it is much older.'

I asked Dorst for his own definition.

'No species ever developed alone', he said 'Each had to become adjusted to the other, like a piece of machinery in a watch or an engine. This is the heart of ecology. I would not say that every biologist is an ecologist, but ecology is not a discipline, it is a way of thinking. An economist, a sociologist, even a philosopher can be an ecologist.' Here Dorst became very animated as he contemplated the possibilities. 'A businessman, an ordinary citizen, a gardener who loves the garden, anybody, can be an ecologist.' Obviously, Dorst felt strongly that ecology wasn't just the preserve of a few dedicated biologists. While this line was not surprising coming from, say Barry Commoner, who was not himself a biologist, it was striking coming from Dorst, who is a distinguished and specialized academic biologist and zoologist from a country whose élitist intellectual tradition rather excludes intra-disciplinary attitudes, and where academics of all types tend to enjoy being a race apart.

'Palaeolithic and neolithic man often had very strong ecological feelings', he went on. 'Primitive people often have them today. I could give you thousands of examples. I have seen the Chagga who live in Tanzania, some of them on the slopes of Mount Kilimanjaro. They are agriculturalists, and they have built up an entirely artificial ecosystem but with such a highly developed ecological feeling, you would just be amazed. They planted different layers of vegetation, so that the soil is well-protected, has no erosion, and produces the highest sustained yield you can imagine.' Then he added, 'Of course, you also have primitive types who do just the reverse and ruin the land. What I mean to say is that you cannot draw a line between ecology and other sciences. I just happened to get interested in this way of thinking very early; and also I had the opportunity early on to visit different countries, remote countries. So by pure chance I got to know some of the simplest, non-European habitats, such as the desert in North Africa and the high altitude habitats in the South American mountains.'

Like Charles Elton and Huxley, Dorst in his early career had found it useful and inspiring to observe at close quarters how a relatively simple environment worked. Although these habitats are the harshest, it seems to be necessary for young ecologists to experience them before turning to the subtler, gentler, and infinitely more complex problems of the temperate zones. Dorst, I learned, grew up mostly in the countryside. 'I was born in eastern France, in Alsace', he told me, 'but my parents did not come from there and we had many relations in the south, so I spent much of my time there. My father was an industrialist, but he was a keen amateur naturalist; in fact he gave me my first bit of equipment, a butterfly net.'

Dorst studied biology and zoology, and was made an assistant at the Paris Museum of Natural History in 1947. He has been based there ever since, making frequent expeditions all over the world. 'I went first to Morocco, to study the adaptation of birds and mammals to desert conditions. I concentrated then on birds – their resistance to drought, their migrations, the adaptations of their reproductive behaviour and so forth. Conditions in the Sahara are peculiar; the dryness, and the great difference in temperature between day and night, and the irregularity of the climate all combine to produce great variations in the migration and movement of populations of the birds and their adaptations. I studied mostly small birds like larks which make excellent material, because they are very abundant and there are a number of different species, each adapted to a different ecological niche.' Dorst still has his students working on these desert birds.

I asked Dorst how the study of ecology stood in France when he began work at the museum just after the war.

'The war itself, I would say, definitely held back the development of the ecological approach in France. We were particularly backward over the way we used land. If we had a tradition, it was oceanography. But there is no basic difference in approach in France, though maybe we arrived at an interest in terrestrial ecology later than some other countries. We have very few true ecologists in France, senior ecologists that is, though in the next generation they will be more numerous. Ecology was not taught at all twenty years ago; now we teach it in most of the universities. But precisely because senior ecologists are so few, we have problems in giving students the specialized studies they need. Often we have to send them abroad: for population dynamics and animal demography we send them to the United States or to Britain.'

Is there a strong conservation movement now building up in France?

'We have several organizations. We have a National Society for the Protection of Nature, which now takes a leading part in conservation battles; and we have a Society which is a conglomeration of local societies, of which there are very many. Most of these are a very harmonious cocktail of all kinds of people, intellectuals, academics, amateurs, and the mixture is proving very useful. There is now a strong amateur naturalist group in France; thirty years ago it hardly existed. When I was a boy there were no natural history books which a small boy or an interested layman could read to know about birds and plants. I remember desperately looking for such books, but nothing could be found. Now I would almost say we have too many – field guides to identify animals, or insects, or mushrooms, and general accounts of different species. And radio and television have done an enormous amount.'

One big battle which these organizations fought recently, with the impassioned support of Dorst himself, was over an insensitive proposal by the French government to use part of one of the finest National Parks in France, the Vanoise National Park in the

Savoie, for a ski resort. The final decision had been taken two weeks before our meeting. 'The President himself declared that the park should be kept as it was, which was a source of great satisfaction to conservationists in France and all over the world', he told me. 'The authorities changed their minds after a huge campaign when public opinion manifested itself in favour of the limits of the park as originally defined. The land under discussion was only just within the limits of the park, but it was a matter of principle. If you declare an area a National Park, it is a National Park forever, as far as "forever" has a meaning for human beings. We could not deconsecrate something that was done only ten years ago, and especially as this fight was fought in what was supposed to be European Conservation Year. This was one of the arguments we used, that France, which wants to have a kind of leadership in Europe, cannot in that very year sacrifice part of a major French National Park.'

At the end of his book *Before Nature Dies* (Collins, 1970) Dorst makes very clear his feelings about the people who argue that land is useless unless it can be developed, and a price of some kind fixed to it. 'The Parthenon serves no useful purpose either; if we tore it down we could erect buildings to shelter inadequately housed populations . . . and yet man, if he took the trouble, could rebuild the Parthenon ten times over. But he will never be able to recreate a single canyon which was formed during thousands of years of patient erosion by sun, wind, and water. He will never reconstitute the innumerable animals of African savannahs, which emerged from an evolution that pursued its winding curves for millions of years before man began to appear among some minute primates.'

Despite his strong feelings, Dorst is awkwardly placed for lobbying. 'I play a part in these battles, definitely, but it is sometimes difficult because I am a civil servant, as a professor in the museum and at the university. Civil servants are technocrats, in a way. We can and must act, and we are paid to give advice to the government; but on the other hand we are not quite free to do everything we can. We are advocate and judge at the same trial.'

One part of France with which Dorst has been particularly concerned is the Camargue, the strange flat marshy region in the delta of the Rhône that is exceptionally rich in bird life. For some time, the Camargue has been protected. The Camargue Reserve was established by the National Society for the Protection of Nature in 1928, with the co-operation of large industrial companies that owned the land. 'It is a paradise for the biologist, with its interesting fauna and the only European colonies of the rose flamingo. As the Camargue's natural habitats range from fresh to salt water, there is nothing comparable in Europe, except for the Marismas of the Guadalquivir in Spain.

'I have know the Camargue since I was a very small boy', he told me, 'and I have seen it change enormously in a few decades. It has become fashionable for holidays, for one thing. Then during the war they started growing rice, which has changed the natural conditions entirely. There are many problems to be solved. In the southern part, where the land is flooded with salt water, we have salt pans, another example of man's impact on the habitat. Obviously there is a conflict between rice-growing, which requires fresh water, and the salt industry, which requires highly concentrated brackish water. Then there is the industrial threat. In the east, at Fos, between Marseilles and the Rhône, they are establishing a huge industrial complex, with a steel mill and heavy industry. This could cause pollution and spoil the habitat, because there are many complex currents and movements of water. Then on the west they have built a vast recreational area, near the beaches, enticing tens of thousands of Europeans to come down and enjoy the sun and the sea. This could cause more destruction than anything else, apart from pesticides which are also a serious problem in the Camargue and linked to tourism, because tourists don't like mosquitoes. The Camargue is a smallish area, and you just can't introduce a large number of tourists. Many people want to see the ducks and flamingoes, but the habitat is fragile, and many of the animals, especially the flamingoes, are terribly shy. One dog walking near a breeding colony of flamingoes can destroy the whole colony, without doing anything directly harmful to the birds.'

Dorst has a special interest in the conservation of wetlands, the technical name for marshes and estuaries. 'The Camargue is interesting not only from the conservation point of view, but also to study how man can best manage delta areas. Wetland ecology is among the most threatened in the world. Marshes are constantly being drained for all kinds of reasons, which is bloody nonsense. For centuries man did, perhaps, have to fight against marshes; there were too many of them, or, they were agriculturally useless, or maybe they were centres of diseases. But the situation has changed; we can now control all these things. And if we suppress all marshes and wetlands in the world, especially those in the temperate zones, it will have a bad effect for the whole balance of natural habitats, because wetlands function like reservoirs: they produce more living material than anywhere else. Only a part of it is directly useful to man, like fish, birds, frogs, and reeds. But the organic material created in these areas is diffused through the earth, and spreads very widely. This is why we must protect marshes and wetlands as long as we can.'

Since 1964, Dorst has been head of the Charles Darwin Foundation for the Galapagos Islands, an international research outfit working closely with the government of Ecuador. These islands are a biologist's dream – a natural laboratory. They have been connected with Darwin since his famous voyage in the 'Beagle' in 1835, when he developed many of his crucial insights into the evolutionary process. 'I have been there already several times, and I hope to go back,' said Dorst. 'It is such a magnificent world. Each island has its own species, often limited to that one island.'

The Galapagos are purely volcanic, he told me. 'The islands came from the bottom of the ocean and they are surrounded by a very deep sea. The soil is all volcanic, and geologically speaking the islands are not very old. It is a beautiful landscape; black, brown, and red lava fields, hugh cactuses and spring plants.' Because the islands have never been connected to the mainland, their animal population arrived, as Dorst describes it, 'by flying, swimming, or on gigantic plant-rafts, like those still carried towards the sea on great tropical rivers. The Galapagos are still in the Reptilian Age, since no mammals have succeeded in reaching them except for one or two rodents, two sea lions, and a bat.'

The story of the Galapagos is a classic instance of the destructive impact of man. Before the Spaniards arrived in the sixteenth century, the islands were untouched. As well as the giant tortoises that give the archipelago its name, there were many other strange reptiles, like land and marine iguanas, and extraordinary and often unique birds, like the Galapagos penguins and the flightless cormorants, birds that stopped growing wings, presumably because there were no predators. When man arrived, Dorst says, the effect was 'immediate and calamitous, especially as the animals showed no fear'. The giant tortoises were slaughtered for their fat, and meat – ten million of them, according to one estimate quoted by Dorst. They have disappeared from some islands and are quite rare on others. As well as these direct attacks, man, by introducing his own domestic animals into the special surroundings of the Galapagos, caused havoc. Pirates brought goats to the islands in the seventeenth century, which bred rapidly and soon competed successfully with the tortoises for food. Then the Spaniards introduced dogs to control the goats but the dogs destroyed the young tortoises and iguanas. Cattle, pigs, rats, and mice all arrived with the invaders, and all had disastrous effects. The pigs dug up tortoise and iguana eggs and the birds also suffered.

Despite these ravages, enough has survived of the unique natural world of the Galapagos to make them possibly the most exciting place in the world for a biologist, especially a man like Dorst who is also a zoologist and ecologist with a powerful interest in evolution. 'It fascinates me. You still have strange creatures like the giant tortoises, and other remnants of very old fauna which disappeared from the mainland during the course of evolution. There are also many specialized animals which evolved there. Some eighty per cent of the birds are peculiar to the Galapagos, like the famous Darwin finches, one of the clues to his theory of evolution. He pointed out quite clearly that they came from the same common ancestor as mainland birds, but became adapted to different ecological niches.'

Dorst has an unusually wide experience of working in different parts of the world and I asked him whether he felt any specially strong attachment to any one part of it, or looked back with special pleasure on any one piece of work.

'My favourite piece of work would probably be the little I did on the ecology of the High Andes in Peru. There are many biological problems there which are of the greatest interest; for instance, how fauna coming from a hot tropical lowland adapted to living in such altitudes. The environment is harsh and difficult. The temperature can rise in one day from –15°C to 20°C. In La Paz in Bolivia, which is about sixteen thousand feet high, one side of the street can be in sun while the other is in shade, and on the sunny side you can't bear to wear your jacket while on the other the ice doesn't melt all day. So the animals have to adapt, and they adapt in various ways, ecological as well as physiological, although at such altitudes oxygen is rare and there are a lot of physiological problems with the blood and the respiratory system. I was also emotionally impressed by the high plateau; by the beauty, by the atmosphere, by the human beings living there. The Indians who live there move slowly. They are not talkative, and they are poor. They are not a happy or extroverted people. They are mysterious. Maybe on other high plateaus it is the same. I have never been, and I shall probably never go, to the High Himalayas. In the High Andes, it's sad; it's monotonous; it can be cold; it's always windy; and it rains; but it has an atmosphere I have not found anywhere else in the world.'

15

David Gates

THE HEARTBEAT OF THE LEAF

MY encounters with the ecological activists, I realized, had led me away from the consideration of ecology as a science. It plainly had another important dimension. It could be an instrument of social change. Especially in America, crusading ecologists were asking us to question many fundamental aspects of the economic and social system. These ecologists, it seemed, wanted to put politics into a new context. Far more attention must be paid, they were saying, to population growth, to the use of resources, to the control of man's material appetites. With Francis Bacon, they were saying that 'Nature, to be commanded, must be obeyed'. Man had been breaking nature's laws and, if he went on breaking them, would soon find himself in such deep trouble that he would never extricate himself. I had also identified the vigilante role of the ecologists: warning us that chemicals now killing birds might also kill people; and that interference with natural systems, whether by clear-cutting in forests or draining marshes, could boomerang on the cutters and drainers.

Now it was time to leave the platforms and the committees and head back into the laboratories. What else might the ecologists teach us in the future? They practised a science that was relatively new, but demonstrably fruitful. Conceivably, it was more relevant to our prospects than any other. In most fields, we seemed not to need *more* science; the problem was to apply intelligently the knowledge we had already. Population control, the most urgent of all problems, didn't need more social science: it was a question of economics and politics, not technology. We knew how to control populations, and we knew how to destroy them. Even in medical science, Sir Macfarlane Burnet, one of the most eminent medical scientists alive, had concluded in his retirement that we had advanced as far as we were likely to go, that the remaining tasks were only to cross the 't's and dot the 'i's. Even in Britain and Europe, ecologists and conservationists were beginning to step up the pressure on governments and corporations.

To complete my picture of the state of ecology today, I decided, I needed to pursue two lines of enquiry. My two initial definitions of ecology, as an abstract way of looking at things, and as a specific branch of biology, needed a third to make the picture complete: ecology as an instrument of social and economic change. I wanted to hear more about that, from younger ecologists who were deep into it. But I was also curious about the future of ecology as a science. As well as the platforms and committees, there had to be the pure researcher in the laboratories. What else might the pure ecologists teach us in the future?

Thinking along these lines, I found myself one day at the Ford Foundation offices in New York. There I heard that they had just handed out nearly half a million dollars to a man who was apparently trying to move ecology into a new phase entirely. His view was that, as a science, it had scarcely begun. So far, it had been descriptive; typically, it described what happened to plants when rabbits disappeared, or observed the relationships between wild orchids and cows. Insofar as ecology had produced general rules, they had been concocted from observation. It had not produced a general theory.

The man being financed by the Ford Foundation had a better idea, he thought. Ecology must move forward from describing *what* happened, and start trying to find out exactly *how* it happened. We must learn to understand nature in a much more precise way. We must discover its basic mechanisms. So far, we had been trying to dominate nature without really understanding it. First, we must acquire this precise understanding, and then we could learn to predict. We would then be able to tell in advance what the consequences of changing the environment would be. We would be able to tell exactly what would happen if we introduced a new pollutant into a particular lake, or imported herds of goats to a stretch of savannah, or used a caterpillar to control a pest instead of spraying it with a pesticide. The name of the man up on this frontier was, I learned, Dr David Gates, Director of the Missouri Botanical Garden in St Louis. He was a physicist turned ecologist, whose work hinged, I was puzzled to hear, 'on the flow of energy in natural systems'. He had more or less created a new school of ecology, known as 'biophysical ecology', and since taking over the Botanical Garden had made it the leading centre in America for training young ecologists in the new approach. Since our meeting, he has left St Louis to become Professor of Botany and Director of the Biological Station at the University of Michigan.

The Missouri Botanical Garden, I discovered, has been a centre of botanical studies for nearly a hundred years. It was founded in 1858 by one of St Louis's original millionaires, Henry Shaw, whose lifelong passion was botany. Shaw was born in Sheffield in 1800, settled in St Louis before he was twenty and built up an enormous business selling English hardware to Indians, fur-traders, and pioneers making their way west. He made so much money that he retired when he was forty, moved to a country house, Tower Grove, and spent the last thirty years of his life developing a Botanical Garden, which he had dreamed of, so the legend goes, since he visited Kew Gardens as a boy. He also started a Library, which now has a collection of 100,000 books, some of them dating back to the fifteenth century, and a Herbarium, with more than two million pressed plants, including samples collected on early expeditions to the West.

Dr Gates has an imposing office in the original Museum building, alongside Henry Shaw's town house, which was transplanted stone by stone from downtown St Louis after his death. An enormous white lily stood in an ornate vase on the desk. Gates is a small, dapper, energetic man of fifty; when I saw him, he had just come from a local TV studio, where he had been on a programme with the Secretary of the Interior, and was smartly dressed in a blue suit, set off by a handsome blue paisley tie, with a red-and-white silk handkerchief prominent in the breast pocket.

Dr Gates is one of the handful of second-generation ecologists. His father, the late Frank Gates, was an ecologist at the University of Michigan, who made a number of pioneering observations of plant life in the woods and lakes of north Michigan, and was a charter member of the Ecological Society of America, founded in 1916; among his friends and colleagues were Aldo Leopold and the great British plant ecologist, Sir Arthur Tansley. 'I spent my childhood in the field with my father; in the Kansas prairie in the winter, and Michigan in the summer. It came as a shock to me when I realized that not everybody knew what ecology was. So, although I was trained in physics and mathematics and chemistry, because of my father I always had a keen interest in living systems. But I could see a big difficulty in the traditional approach to ecology, and my father agreed with me. While it was still in a descriptive stage, you couldn't make it a predictive science. My early work was in atmospheric physics studying radiation, water vapour content, and meteorology. Then I decided to try to apply physics to ecology. I wanted to try to understand how organisms actually operate in the environment. We had to find a precise way to understand how plants and animals are coupled to the environment. The way they are coupled is through energy. All life, everything we do, every process of life itself consumes energy. If we are to understand how wind or moisture affect a plant or animal, we can only do so through understanding the flow of energy.

'To understand an ecosystem, which may be a lake, a forest, a meadow, a mountain-top, or a bog, you must understand the two basic components: the biotic – that is, the living organisms it contains; and the abiotic – the physical processes within it. If, as scientists working with these ecosystems, which are complex, we work only as biologists, we shall not succeed. And if we work only as physical scientists and look at the abiotic side only, we shall not succeed. The only way you can do ecology well is by using the full breadth of science.'

I remarked that this seemed rather a tall order. Dr Gates became indignant. 'It's a hard science but we have no choice. To be a true ecologist you have to understand some of the biochemistry, the molecular biology, genetics, evolutionary characteristics, community structure, meteorology, physiology . . .'. Dazed, I suggested that many people who felt they had recently 'discovered' ecology were not aware of all this.

'People confuse environmentalism with ecology', said Gates firmly. 'They don't know what ecology really means. And I can tell you that some of the people who think they are ecologists don't know either. It's one thing to take a qualitative look at the thing, it's another to really take its pulse. Ecology must be developed to as analytical a stage as possible. The trouble is that in any ecosystem there is such a large number of variables: radiation, air temperature, wind, moisture, rainfall, soil properties, nutrients, plant and animal properties, heartbeat, blood circulation, fat, fur, colouration – all these are parts of the system. So how do you navigate? By mathematics. If you use the analytical properties of mathematics, and formulate the problems properly, you can put a lot of these properties together simultaneously. You take the flow of energy as the common denominator, and you use good mathematical formulations based on good mathematical principles.'

But before you can apply mathematics, I suggested, was it not essential to make sure that your basic observations of ecosystems were complete?

Gates insisted that the two methods must work side by side. 'It's the only way you can make ecology a predictive science, which is what we need', he said. 'The descriptive phase of ecological work is good in itself, and necessary, but it doesn't get us to any precise definition of why something works the way it does, and that's what we must discover. Take a leaf. How does a green leaf function? This is as fundamental a question as any faced by man. Everything begins there with the primary productivity of green plants. First, you have to have energy to drive the system. The green leaf breathes in carbon dioxide through its pores. We have to understand what happens inside the leaf after that. There's gas diffusion, which gets the CO_2 in and drives the oxygen and water vapour out, and there's the metabolism of the leaf, how it functions in assimilating carbon dioxide and making carbohydrates. That gets you into biochemistry. We're trying to take the heartbeat of the leaf. If we can't understand how a plant functions, we aren't going to get very far with the management of the world.'

Much of the work by Gates and his colleagues in the research laboratories of the Botanical Garden has been on the properties of leaves. They have produced an exceedingly daunting equation for the energy exchange of a plant leaf. No amateur ecologist, plainly, is going to be able to follow much, if any, of the detail of Gates's work. He is also doing the same kind of study with animals. 'We want to know how the animal lives, what are its energetics and functions', he told me. 'Then we'll want to know how animals live on plants, and how other animals live on animals that live on plants, and so on.' Already, he said, they have reached a stage where they can predict the climate that certain animals and plants must have if they are to function properly.

For Gates, the future of ecology is challenging, but bright. 'In the future we will see ecology done very much better than it is today', he said. 'It's still a primitive science. It will go way beyond its present standards of sophistication and elegance and intellectual rigour. I think ecology is going to be the complete science.' Here Gates became almost frantic as he tried to convey his vision. 'Do you see what it means?' he demanded. 'Do you see what it *means* when I try to tell you that ecology means understanding the biochemistry and the enzyme systems in each organism; understanding why they're unique, and why they respond the way they do to all the variables of temperature, and moisture, and alkalinity, and acidity; and how you have to understand the cell, and its structure, and the breathing of gases, and the taking up of CO_2, and the giving out of oxygen, and the play of energy in the system, and the whole environment? You see what an enormous, beautiful piece of science this is?' He seemed exalted as he contemplated this dream: but reality broke in. 'The level at which this can be done is so far beyond what we've yet done that to future ecologists we will look pathetic, like a bunch of hacks.'

Although Gates obviously thinks that most of the people who have recently discovered ecology have no idea of its true complexity, he has contributed himself to what he calls environmentalism, and acknowledges the importance of the environmental movement. In the last few years, he has served on government advisory bodies, and done his share of campaigning against pollution and the waste of resources. One of his special themes has been the importance of maintaining diversity in natural systems, because diversity creates stability. 'In agriculture, man replaces diversity with monoculture, which means he has very vulnerable and unstable crops', he said. 'But, at least people are now realizing that if they disrupt something at the bottom of the pyramid, the whole structure may be in danger.'

What Gates was after, I decided, was recognition of ecology on two levels; in general terms by the public, and in tough scientific terms by the professionals.

There have been three significant breakthroughs, he told me. 'First, people have quite recently begun to realize that the earth is one system, and we need ecological understanding of it if we are to avoid disaster. Secondly, the advent of the large computer, in the fifties – this was crucial, we couldn't do modern ecology without it.

It is essential for data-handling when complex systems are involved. Thirdly, we are beginning to get a new generation of ecologists trained in mathematics. But it's a long process. I reckon it takes eight to ten years to turn out a well-equipped ecologist. A lot of the biologists don't have the analytical mind needed to use mathematics, and a lot of the physicists and mathematicians don't have the biology.'

Feeling rather overwhelmed by Dr Gates's vision of the new breed of ecologist, I went off to watch one of them in action. The research laboratory turned out to be a reassuringly ramshackle group of low buildings containing benches littered with plants, one quite small computer, and a number of amiable young men in shirt sleeves looking more like casual gardeners than super-scientists.

I talked to a former student of Dr Gates called Elwynn Taylor, who looked about eighteen but had just taken his doctorate and so could be considered a fully-fledged ecologist.

Taylor had spent four years, he told me, studying the problems faced by leaves. How much heat can they stand? How do they use rain? He had discovered that a small tree native to Missouri, the redbud, adapts to its environment by changing the position of its leaves. It does not like too much sun, so in fierce sunlight the leaves hang vertically, which means that a much smaller part of them is exposed to the sun and their temperature is thereby kept low. Meantime, the plant takes in the heat and light it needs by keeping its leaves horizontal on the shady side away from the sun. He showed me a series of photographs of leaves in different positions.

'I began to make predictions, based on Dr Gates's mathematical models, about how the size of leaves relates to their ability to survive in different climates', he said. 'I planned to spend a summer researching the question in Panama, so before I went I made a series of predictions about what size of leaves should be suitable for the climate there. My predictions said that all leaves over ten centimetres wide would be in trouble.' We have always known from observation, he explained, that plants in dry climates have small leaves, whereas plants in wet climates have large leaves. He was trying to establish exact parameters. 'But I was quite distressed, because I knew bananas grew in Panama, and I knew that their leaves were wider than ten centimetres. When I got to Panama I discovered that the banana plants there are nearly all quite tattered, which means each individual strip of leaf is small enough not to overheat.' This had made him wonder about palm trees, and whether their long thin strips of leaves might not be a permanent version of the tattered banana. 'There may be some evolutionary aspect here, but we haven't looked into it yet.'

His next piece of work, he told me, was to be an investigation in detail of how certain plants seem able to adapt instinctively to different atmospheric and meteorological conditions. It appears that some plants, because of their location, have had to develop much more flexibility of response than others; the redbud, for example, is able to cool itself by increasing its rate of evaporation as and when necessary. 'Almost', said Taylor, 'as if the plant could make a decision'.

Apart from the pleasure it gave him to discover how these processes worked, this knowledge could be very useful to plant breeders and agriculturists. 'If I can find out', he said, 'how widespread this sort of adaptation is in nature, and how it works, maybe that capability can be bred into agricultural plants for use in semi-desert conditions.'

I remarked that this work, fascinating and useful though it was, had little to do with fashionable environmental agitation. Taylor laughed. 'I guess you know that Dr Gates makes a distinction between ecology and environmentalism: I agree with him on that. My work has nothing to do with pollution, though there are people here doing research on how plants are affected by chemicals, for instance, automobile exhausts. But at a personal level I try to be responsible towards the environment, though not radical. I still own an automobile.'

He took me round the laboratory, part of it a converted basement where, not long ago, the Garden used to grow mushrooms. He showed me a wind tunnel, where mice and small birds were tested for their reactions to wind, temperature, and light. 'It's easier to experiment with leaves', he remarked. 'Less hurtful.'

16

Edward Wilson

LEARNING TO PREDICT

AFTER I left St Louis, I discovered that a number of younger ecologists have for some years been applying the mathematical approach propounded to me by Dr Gates. One of them, Dr Edward Wilson, is a professor in the Zoology Department at Harvard, and I found him in the Biology Laboratories building, an imposing redbrick structure with two large stone rhinoceroses flanking the entrance.

Wilson, a thin, dark man in his mid-forties, started out as an entomologist with a special interest in ants. The walls of his narrow, functional office were decorated with large photographs of ants; one, in colour, was especially striking, showing an ant-like insect in close-up through a weird orange haze. This creature, he told me, was the oldest ant ever found, dating back a hundred million years; it had turned up in a piece of fossilized amber chipped off the New Jersey cliffs by a couple of amateur rock-collectors in 1967. 'It was a great moment', he said. 'No one was *trying* to find the ancestor of the modern ant. Some of the greatest pleasures of science are caused by the sudden occurrence of the unexpected.'

As well as the physical and social evolution of the ant, Wilson has worked on all kinds of problems in insect ecology, including population structure, distribution, and adaptive characteristics. He immediately confirmed what Dr Gates had told me about mathematics and modern ecology. 'What we are seeing is the gradual development of ecology into a hard science', he said. 'Previously, much of ecology was descriptive. Where general rules were formulated, they were generalizations from specific observations; there was very little actual theory. But nowadays younger ecologists feel that if we are going to extend ecology as a hard science, we must make it predictive, so that we can tell in advance what the consequences for the environment will be when we add a pollutant or introduce a new species. This approach is not yet by any means universally accepted as the wave of the future; but it is emerging as an attempt to make ecology more scientific.'

Were some ecologists then resisting this tendency?

Wilson chose his words carefully. 'Let's say it has been a matter of some dispute between older ecologists and some of the younger school over the last ten years or so.'

Standing behind the younger men, of Wilson's age or younger, are a handful of senior men, he told me, who began twenty to thirty years ago to propel modern ecology away from naturalist observations and towards mathematical theory. He named in particular Gates, Eugene Odum, and G. E. Hutchinson. Odum, the author of the standard college textbook, *Fundamentals of Ecology* (W. & B. Saunders, 1959) is an ecologist at the University of Georgia who has worked on a variety of systems from estuarine marshes to upland forests: he, like Gates, started early on to build up a body of theory about the flow of energy through ecosystems. Hutchinson, an Englishman by birth, has been teaching ecology at Yale for twenty years. 'His influence should be heavily stressed', said Wilson. 'He was a top descriptive man, specializing in lakes and rivers, who was an avowed ecologist when it was still a very unfashionable subject. He took up mathematical theory early on, and has turned out a high proportion of the best younger ecologists working on this approach today.'

One of these, Robert MacArthur, now at Princeton, gave the burgeoning mathematical school a sharp shove forward with a series of papers, starting in 1955. 'MacArthur, an old friend and colleague of mine, is a very significant and controversial figure', said Wilson, adding hastily, 'if you happen to meet him, don't tell him I said so, because he probably already gets more praise than is good for him. But seriously, MacArthur, who trained as a pure mathematician before he got into ecology, came right out and started tackling complex problems, such as what controls the numbers of species, on a simple theoretical basis. This caused quite a stir. Some senior ecologists – no, I won't name any names – accused him of being facile, of grossly oversimplifying, and insisted that ecology is too complex and delicate to be approached in this way.'

Wilson himself, however, very quickly became convinced that this was the right direction and started to work with MacArthur to see how his philosophy of theory-making could be developed and applied, in particular to island populations. 'Biologists like MacArthur and myself, and other scientists at Harvard, Princeton, and the University of Chicago especially, believe in what has come to be called "simple theory", that is, we deliberately try to simplify the natural universe in order to produce mathematical principles. We think this is the most creative way to develop workable theories. We don't even try to take all the possible factors in a particular situation into account, such as sudden changes of weather or the effects of unusual tides. Our notion is to proceed rapidly and intuitively to get at the most important and universal factors.'

For several years, MacArthur and Wilson laboured together to produce an equation that defined the point at which populations of plants and animals on an island might be expected to reach an equilibrium. This highly technical exercise involved stating the number of species that might be found on the island, the rates of arrival of new species and the extinction of others, the rate at which colonization would take place, and the survival rates of the species. Then, like any scientist with a new theory, Wilson set about thinking how it could be tested. He worked out, with another colleague from Harvard, Daniel Simberloff, one of the first experiments attempted with an entire natural ecosystem.

They went down to the Florida Keys, that strange area where the southern tip of America trails off in a long archipelago of small, flat islands linked by U.S. Highway 1. There, amid palm trees, hurricane warnings, and tourists going deep-sea fishing, where the presiding spirits are those of Cuban gun-runners and Hemingway's social exiles, they picked out six extremely small mangrove clumps in the shallow waters of Florida Bay. These clumps would look to most people like bushes sticking out of the sea, but, as Wilson explained to me, a piece of mud with a bush on it a few feet across is as much an island to an insect as Cuba or Great Britain is to a migrating bird.

After they had selected six islands, which had to be near enough to the larger islands or the mainland for insects to be able to reach them in large numbers, they made a careful survey of all the resident insects. They found that about seventy-five insect species usually inhabited their islands. Many birds, including green herons and white-crowned pigeon, visit the mangrove bushes, and water snakes and the occasional raccoon make use of them from time to time, but Wilson and Simberloff were concerned only with insects.

They then set about systematically removing all the insects from the islands. It took them some time to find the best method. Chemical spraying was not 100 per cent effective, so they decided to try fumigation, which involved erecting a reasonably airtight covering and puffing in gas. They tried various kinds of gas, because they needed one that was not soluble in water and would not harm the vegetation. They found that if they fumigated during daylight, even when the sun wasn't out (which down there it usually is), the extra heat severely damaged the mangroves, so in the end they decided to do all the fumigation at night.

They then were faced with what Wilson describes as, 'The overwhelming physical problem of how to fumigate an entire island'. If they simply draped their tent, which was made of plastic-impregnated canvas, over the bushes, its weight broke the branches. They needed some kind of superstructure that would keep the canvas off the bushes and would also be collapsible and re-usable; it would have been too expensive to build a permanent dome that could be lowered over each island in turn by helicopter.

They fumigated the first island by using an elaborate scaffolding, which proved effective but clumsy. They eventually compromised with a steel tent-pole, erected high enough to take the weight of the canvas, which was winched up to the top of the pole and then carefully lowered and secured with sandbags and stakes below the water. Wilson showed me pictures of the fumigation tent in position; it looked as if a partly-submerged balloon was floating on the surface of the sea.

They then killed off all the bugs and flies on the islands. To make absolutely certain that there were no survivors, they broke and examined literally thousands of hollow twigs, lifted or peeled off loose bits of dead bark, squirted mild insecticides beneath bark and into hollows to drive out stragglers, and dug into likely hiding places with a steel probe. All this work had to be done with great care to avoid damaging the insect habitats, which would have affected recolonization.

For Wilson and Simberloff, the excitement was now beginning. For nearly nine months, they visited the islands every eighteen days, minutely observing the recolonization process. Each time, they went through the twig-breaking, bark lifting, tapping routine. To ensure that they brought no insects with them when they visited the islands, they usually anchored their boats some way away, scrupulously examined their bodies and clothes for any lurking creatures, and regularly sprayed themselves and their equipment.

The insects began to return. Wilson and Simberloff noted the order in which the various insects arrived; the rate at which they settled and began to breed; and the method by which they made the journey. 'Invasion', as the process is technically called, took place in a number of ways; the insects flew, swam, were carried by winds, and arrived via birds, as parasites, or concealed in twigs. Some came drifting in on rafts of floating vegetation. The air, which most people think of as 'empty', is in fact thick with invisible or barely visible forms of insect life, 'aerial plankton' as one early entomologist called it. Millions of these creatures circulate without finding a resting place and eventually die off.

After two hundred and fifty days, they found that all the islands except the most isolated one had been recolonized by almost exactly as many different species, and in much the same proportions, as before – although the total number of insects was still relatively low.

The findings confirmed the theory they had set out to test, that 'a dynamic equilibrium number of species exists for any island'. This process occurs in three stages: the insects arrive and settle; then they start to interact, by breeding, competing for food, and eating each other; finally, they reach a state of natural balance.

Wilson was plainly extremely gratified at the success of the project, which had occupied him for nearly three years. 'I suppose it did create a fair amount of excitement', he said modestly. With MacArthur and Simberloff, he had succeeded in pioneering a mathematical way of predicting a natural process.

The story exactly shows the crucial difference between the descriptive and the experimental approach. Old-fashioned naturalists' methods remain an important part of modern ecology; there was plenty of basic observation involved, but it was directed to a particular end. Besides, an experiment like this requires a rather more detached attitude to the wonders of nature than is perhaps usual among ecologists, and Wilson, I discovered, had felt the need to protect himself against criticism. When presenting a paper to a group of his colleagues, he added a revealing postscript directed at 'Conservationists who, very understandably, will be concerned with our concept of doing ecological experiments.' He explained that he and Simberloff chose their six experimental islands from among hundreds, after consulting biologists and conservation officers from the Everglades National Park and the Great White Heron National Refuge. They made every effort not to damage trees or local marine life, though they did kill off all the trees on one of the islands during a trial run. 'Such effects', he said, 'can be eliminated altogether by improvement of the defaunation technique in future experiments . . . Two years after the experiment was performed, the little islands and their surrounding environments are essentially as we found them. I believe in the principle that such experiments should be designed with extreme care. In no case should ecologists, of all people, be guilty of causing significant irreversible harm to the environment.'

Next, I asked Wilson whether he, like Dr Gates, saw ecology gradually becoming the complete science. He frowned slightly. 'Modern ecology certainly requires the most diverse scientific talents', he said, 'and it is a highly eclectic subject. All kinds of new components are being fed into it.' As an example, he told me of another branch of the science which he has studied: chemical ecology. 'We can't understand the behaviour of many animals without understanding the chemical communication that goes on between them.'

He showed me a paper he had written on 'pheromones', as the chemical substances secreted by animals are called. He has studied the ways in which these chemicals are used by moths and ants to attract or repel other insects. 'The ecological implications of these findings are considerable', he wrote. 'We know that most animals in a given ecosystem are being guided from moment to moment through the mediation, at least in part, of chemical signals.' As he explained to me, this field of study has been neglected partly because man is himself so dependent on sight and sound for communication that the idea of communicating via chemical secretions seems especially obscure. Certainly, the analysis of an ant gland was the smallest scale and most bizarre ecological enquiry I had encountered.

But Wilson turned out to have yet another new field of ecology up his sleeve. 'Then there is behavioural ecology', he said. 'You could define ecology as the study of the behavioural patterns of animals in their natural surroundings. We need to understand the ecological significance of behaviour patterns, and this is where ecology links up with ethology, isn't it?' I said I associated ethology with the recent appearance of popular books seeking to explain human nature in terms of animal behaviour, like Desmond Morris's *The Naked Ape* (Cape, 1967) and Robert Ardrey's *Territorial Imperative* (Collins, 1967). Wilson, it turned out, had recently written a paper criticizing such comparatively simple analogies. In his view, territorial and aggressive 'instincts' in animals and, indeed, in man, are not simply inherited or built in, but are adaptive reactions to particular circumstances, such as unusual stress or high population density.

But although Wilson sees ecology as the focus for a number of overlapping new subjects, he does not regard it as all-embracing. Before I left, he outlined for me what he called 'a subcontroversy' among the younger ecologists themselves. 'On the one hand you've got the hard ecologists like MacArthur and myself, who, as I've explained, believe in simplifying theory as much as possible. You can call us the simple theorists. But in the last five years or so a group has developed, around people like Paul Ehrlich at Stanford, C. S. Holling at British Columbia in Canada, and Kenneth Watt at Davis, who are also mathematical ecologists, but who believe in complex theory. That is, they believe ecology can save the world. So to them, everything is relevant. These guys have a political argument and an intellectual argument. Their political argument is that all ecologists should harness themselves to the movement of applied ecology, plotting the management of the world's fisheries, re-routing water systems, managing the world's forests and so on. To me, these are social engineering problems, and there's not much ecology in them. There are basic ecological *principles* involved, but that's all. Their intellectual argument is, in my view, even weaker. They say that because ecosystems are so vastly complex, you must be able to take all the various components into account. You really must feed in a lot of the stuff that we simple theorists leave out, like sunsets and tides and temperature variations in winter, and the only way you can do this is with a computer.

'To them, in other words, the ideal modern ecologist is a computer technologist, who scans the whole environment, feeds all the relevant information into a computer, and uses the computer to simulate problems and make projections into the future. Sometimes they seem to believe that this is the only way to operate, and occasionally they accuse people like me of wasting our time, playing with ideas and hiding in our ivory towers while the world is coming to an end.' Wilson became indignant. 'What the complex theorists have done up till now', he went on, 'is to tell us how things should be done. I know I'm biased here, but in fact it's been mostly the simple theorists who have made new discoveries about how nature actually works. The complex approach may well pay off in the end, but it hasn't yet produced many discoveries of general interest. Their main achievement, in my view, has been to alert the public about environmental problems. And this has been tremendously valuable. I want to give them all credit for it. It can only be good for ecology. Someone said not long ago, "Pollution will do for the study of ecology what cancer did for molecular biology". Public interest means more money for research and better young scientists wanting to be ecologists.'

Once again, I decided, the accident of individual temperament was proved to be significant. Although with Wilson I had talked mainly about his academic studies, he had established early on that this specialized work was essential if man was to learn how to measure and control his impact on nature. It was not that Wilson didn't care about the broad implications of his work, or that he took no interest in the environmental crusade – he had been one of the original members of Z.P.G., Ehrlich's population control organization, and he has been active in the conservation movement in South Florida. But for him, what mattered was the substance of ecology itself, and the struggle to make it more, not less, scientific. His energies were all directed that way and there was not much time left over for campaigning, or for thinking in global terms.

17

Kenneth Watt

THE LYNX AND THE COMPUTER

FROM THE time I first started looking into ecology, Kenneth Watt's name had cropped up repeatedly in connection with the future of the science. Everyone in ecological circles knew that he was working with computers on an extremely ambitious study of global environmental problems, but the whole thing seemed a bit vague. The general impression I got was that Watt, who had started off with a base in the Canadian Department of Forestry, where he used computers to study pest control problems, had flipped out not long after taking up a job in the University of California's agriculture offshoot at Davis, near Sacramento, and had become a sort of prototype of the eccentric scientist brooding on how to set the world to rights. When I found myself in California, I decided to pay him a visit.

My first contact with Watt did nothing to dispel my intimations of his eccentricity. Over the telephone, he agreed to talk to me, then added in a manic tone, 'But I keep a bear in my office, so you'd better wear stout boots'. He then asked whether I would expect him to talk about himself and when I said 'Yes', he informed me that I might not get much out of him, because his mother had been a lizard and consequently he had no navel.

After a long bus-ride up to Davis round the edge of San Francisco Bay (through what must have been one of the most beautiful environments in America before man and technology got at it) I arrived at the Davis campus not quite knowing what to expect. I found Watt wearing an ancient green sweater in a small, bare office off a corridor leading to a room full of filing cabinets and computers. He is a tall, shaggy, bespectacled man in his early forties; with longish greying hair; he seemed to be a compulsive talker, his words tumbled out in a voice that frequently slid up the scale into a squeak of excitement or exasperation.

First, I felt the need to establish what he was actually doing with his computers. I learnt that he is head of a team called the Environmental Systems Group, which is part of a new Institute of Ecology at Davis where they are constructing a mathematical model of the State of California. Their aim then is to conduct computer simulation studies to elucidate the effects of the state's steadily rising population. Already this seemed less wild than I had imagined; and it seemed less wild still when I learned that the money behind this project, which will take years to complete, had been put up mainly by the Ford Foundation as part of its investment in important ecological research.

The scope of the study is tremendous, as I discovered when I looked through the progress report handed to me by Watt. Entitled 'A Model of Society', it stated plainly that California was chosen as the basis for a further study of the whole of human society: 'The time is overdue for an attempt to synthesize information from a variety of disciplines into a form capable of analysis as a quantitative, albeit probabilistic system . . . human society, with its complex environmental dependencies and manipulations, recognizes no arbitrary intellectual boundaries.'

I found, over the four hours or so I spent talking to Watt, that he doesn't recognize boundaries much himself. He stated his basic theme at the beginning of the conversation: 'Our society is breaking down in several senses. All technological societies, capitalist and Marxist alike, have certain built-in defects. And what have been perceived as the real problems so far are, in fact, only symptoms. What we're doing is to identify the real problems and offer solutions.' After this, he talked more or less at random, giving examples, often at length, from the California study, his earlier work, and his personal impressions of everyday life.

He launched into the population problem. This is something he is very much concerned with and Paul Ehrlich is a close friend. The computer study, after all, is trying to estimate what happens to all aspects of a society when the population keeps going up. California, where the rise has been particularly abrupt and dramatic, made an excellent model for the rest of the world where the rise is expected to be equally dramatic in the next two or three decades.

Watt has unearthed some interesting and little-known effects of population growth. 'If a population is constantly going up', he said, 'you increase the numbers of young people in that society. In other words, you increase the number of people in the education-tax *consuming* age group in relation to the education-tax *producing* age groups. That is, a comparatively small number of taxpayers are going to be paying for the education for an increasingly large number of kids. And they aren't going to like it. Our group has done a number of pioneering studies on this aspect of population growth. It's just one of the ways you can demonstrate that population growth isn't even economically a good thing; that it doesn't bring down taxes, which is what some people still think. We've shown that a 1 per cent rise in population can lead to a 25 per cent increase in the educational-tax burden per tax payer in a given community. In a place like San José, where the population has increased 114 per cent in the last ten years, they're obviously going to be wiped out by taxes. Of course, this rising taxation leads to social tension – older people grumbling about ungrateful long haired kids, why should they have advantages we never had, and so on. This isn't just an irrational attitude; there is a real social problem underlying that kind of grievance.'

He went on to say that he thought rising populations had a good deal to do with the generation gap in general. 'There are more young people in relation to older people than ever before in this country, which is why we're having so much difficulty in transmitting our traditional cultural values.'

The California study covers, as well as recognizably environmental problems like air and water pollution, and land use, all kinds of urgent social problems: racial conflict, crime and delinquency, and decision-making. How decisions are, or are not, made, has preoccupied Watt a good deal lately and he has written a book on the subject. 'There's no planning worthy of the name being done in this country or anywhere else', he said, 'and our decision-making processes are totally inadequate. For a start, one of the most important options in any decision is usually ignored, the option to decide to do nothing at all. And communication between the different hierarchies in our large organizations is almost non-existent. They're organized along parallel lines, like this.' He seized a pad, drew a diagram and handed it over. It was a diagram showing the basic construction of Hitler's inner council.

How could you make rational decisions without effective communication?

Rather than risk getting snarled up in organizational traps, Watt inclines to the direct approach. For instance, he had a suggestion he wanted to put to President Nixon about how to help the big aircraft companies out of the mess they are in. 'Nixon is committed to giving millions of the taxpayers' money to the big companies. Now the commercial airlines are in terrible trouble, because they completely miscalculated public demand. Anyway, what we need is not more aircraft, or more cars for that matter, since the air and the roads are already overcrowded, but good, cheap, rapid transportation systems. Why doesn't Nixon give companies like Boeing the two hundred and ninety million dollars that were earmarked for the cancelled S.S.T. and tell them to develop highspeed trains like the one in Japan? What's wrong with the idea? Is it evil? Who loses by it?'

At some point I managed to break in and ask Watt how he had progressed from insect problems to freewheeling social engineering.

'It was the volcanoes', he said. 'I'd always had an intense interest in biology, as I still have. I love to have, say, raccoons as pets. When I was a student at the University of Toronto, studying grasshopper populations, I began to get into calculus and statistics, and it was then that the light dawned about how all this stuff could be applied to real problems. Once the light dawns, you're never the same again. First, I concentrated on the population dynamics of fish and insects, then in the early sixties I started to work on social and environmental problems, connected with resource management.'

So where did the volcanoes come in?

'Oh, yeah, the volcanoes. Well, I began to be interested in epidemiology, the study of plague, and how it connected up with global weather patterns. Then I realized – it was suggested to me by a colleague, actually, Dr David Deamer – that one way to study this was to trace the effect of volcanic eruptions on climate. This led me to look at environmental disruptions on a global scale.'

Watt has unearthed some astonishing correlations between volcanic eruptions and distant events. 'By studying *Duffy's Farm Journal,* a nineteenth-century British newspaper, I've been able to relate fluctuations in the London and Liverpool commodities market with volcanic activity in South East Asia', he said. Apparently it is well-known in meteorological circles that for some years after a major volcanic eruption the world weather is significantly cooled down, as a result of the vast masses of 'fine particulate matter' hurled into the atmosphere up to sixty miles high, which affects crops, and hence commodity markets. It also effects human and animal health. Later, I came upon a paper of Watt's which indicated an extraordinary similarity between fluctuations in the number of lynx in the Canadian Arctic and the insidence of influenza outbreaks in England and Wales between 1860 and 1950. Watt suggested that both the high points of lynx mortality and the flu epidemics could have resulted from cold weather caused by volcanic eruptions like Krakatoa in 1883, and Katmai, in Alaska, in 1912.

Another recent preoccupation, Watt told me, has been 'the deteriorating relationship between men and women in our society'. He relates this directly to man's callous treatment of nature. 'My wife and I were in Fiji this summer. Fiji is my Valhalla; I'd happily spend the rest of my life there watching fish among the coral reefs. We happened to meet a man in Fiji representing some American real estate development firm. We noticed that as well as having a very brutal, exploitative attitude to the island, he seemed to have the same kind of attitude to people. He really had the rapist mentality; and if someone really doesn't care about people, how in God's name can you expect him to care about ruining a beach or a forest?' Watt sighed, then added, with a note of regret, 'unfortunately, it's extraordinarily difficult to look at these things in a scientific fashion.'

I left Davis with my head spinning, but at the same time impressed by Watt's energy, enthusiasm, and apparently unshakeable confidence in the capacity of men, with the help of the computer, to get a grip on their most intractable social and environmental problems. He and his team were pushing way ahead of ecology, into complex social mechanics, trying to map the future; trying, by letting their computers loose, to show man that he did have choices. It seemed especially suitable that this work was going on in California, which has moved so rapidly in the last fifty years in the direction that much of the rest of the world seems to be trying to follow. In a preliminary statement of what the Environmental Systems Group is trying to achieve, Watt wrote: 'The computer will not make decisions for the future. People must. But without the kind of information the computer can yield, the comprehensive programs we need to meet the environmental challenges ahead will not be found. Our next step would clearly seem to be to hitch the immensely valuable potential of computer analysis fully to the planning function. This has not yet been done. The state should forcefully take the lead to this end. It is a matter of life or death, and what California chooses, so can the world.'

18

John Milton

ECOLOGY AND TOMORROW'S WORLD

STIMULATED THOUGH I was by the frontier activities of Wilson and Watt, I felt a powerful urge afterwards to talk to someone directly involved in an immediate, practical application of the subject. So I went to talk to John Milton at the Conservation Foundation offices in Washington. His central concern, I had heard, was to get ecological knowledge taken seriously in the less developed countries of the world.

Milton is thirty-five, and looks more like a rising young film director than an ecologist; he dresses fashionably and has a neatly drooping moustache. His official title in the Foundation is Acting-Director of International Programmes. He reckons he started practical conservation work at the age of thirteen. When one of the few remaining wilderness areas in New Jersey came up for development, he organized a guerilla group of his friends to pull up surveyors stakes and put sugar in the bulldozer's petrol tanks. 'We held them up for about a year and a half', he told me.

He went on to study conservation and ecology at the University of Michigan under Stanley Cain, an outstanding educator as well as a distinguished plant ecologist and once Assistant Secretary of the Department of the Interior. Milton joined the Conservation Foundation straight out of college and has been with them ever since, travelling constantly as an adviser on environmental management and protection in Africa, Latin America, and Asia. He has helped to organize several conferences, including the important one on Future Environments of North America, and written numerous books and papers.

His interest in the less-developed countries dates back to before his university days, when at twenty he went down to the Yucatan peninsula in Mexico and, as he put it, 'By some fluke happened to discover a large, unknown Mayan city in the jungle'. He was given a job as an archaeological expedition guide in Yucatan by the Mexican government, and became fascinated by the Mayan Indians he encountered, who were 'quite uncontaminated by Western culture'. He then spent a year wandering between Mexico City and Panama, studying cloud forests and acquiring an admiration, which has endured, for primitive cultures and balanced ecosystems.

For his master's thesis, he undertook a study of a Development Institute in Costa Rica, at that time virtually the only place in Latin America that was attempting to train specialists in land resource problems in tropical areas. 'The thing that has bugged me almost more than anything ever since', he said, 'is that almost no training in ecology is going on in the Third World. Nearly all the training in the developed countries is oriented towards temperate zones, and when ecologists trained in this way go to work in tropical countries it takes them two or three years before they begin to understand tropical systems. In that time, they may well be making serious mistakes. What we should be doing is to train Latin Americans, and Asians, and Africans, who after all represent two-thirds of the world's surface and three-quarters of the world's population, to look into their own special problems.'

Milton sees himself as having a precise mission: to make development agencies, both American and international, pay more attention to the ecologists. 'The World Bank, AID, and the U.N. agencies are just going to have to get into dialogue with ecologists', he said. He is in a pretty good position to bring influence to bear on the World Bank, the world's major investor in development projects; one of his research assistants at the Conservation Foundation is Kathy MacNamara, whose father, Robert MacNamara, former Secretary of Defence under Kennedy and Johnson, is now head of the Bank. 'Macnamara really does have an environmental frame of reference, more than any other head of an international aid organization', said Milton. 'It's quite exciting the way the Bank is beginning to develop a body of ecological expertise.' Just before I talked to Milton, the World Bank had belatedly appointed their first ecological adviser.

Milton has strong views about the attitude of developed countries to the Third World. 'We ignore their special problems', he said, 'which is completely irresponsible. The rich western countries have got rich at the expense of the Third World, and implanted western assumptions and desires. We are still exploiting their resources and agriculture and cheap labour. As I see it, the world situation is analogous to the racial situation here in America. The blacks were an exploited agricultural minority in the South, first as slaves and then as tenant farmers. They were kept at the lowest economic and social level, and given no chance to acquire any wealth. They began to move into the cities and a new kind of exploitation developed along with the urban ghettos. The blacks in this country have been kept down and used and used and used. Now we've got a revolution on our hands. But in this country, the blacks are a minority; in the world as a whole the less-developed peoples are in the majority, and they have most of the world's resources. If we don't change our ways, there's going to be chaos.'

Even when the rich countries do invest in development programmes in the Third World, they frequently do so in the wrong way and for the wrong reasons. 'Most forms of aid are designed to help our economies, not theirs', said Milton flatly. 'We export our own technology because it's cheaper and easier to export what we've already got than it is to work out new technologies adapted to what other countries really need. That's why so many of our projects have disastrous effects abroad – particularly in the tropics and arid lands.'

Milton and a number of other ecologists became more and more alarmed by this pattern. They felt that development agencies were not taking ecological criteria into account when their schemes were in the planning stage, and also that little was being done to evaluate existing projects to learn from some of the more flagrant mistakes. Admittedly some individual ecologists, like Frank Fraser Darling and E. B. Worthington of Britain, were trying to direct attention to those mistakes, but there was no concerted effort being made.

So, in 1968, largely at Milton's urging, the Conservation Foundation joined with Barry Commoner's Centre for the Biology of Natural Systems at St Louis in sponsoring the first big study, followed by an international conference, on Ecology and International assistance. The research and the conference involved experts from all over the world, plus representatives of all the major development agencies, and discussion was focused on a series of case histories that make absorbing, if depressing, reading.

Soon after the war, in the Canete Valley of Peru, American

agencies promoted the widespread use of pesticides against insects damaging cotton. Seven years after the treatment started, the cotton crop was down by half and the number of destructive insects had doubled.

The Aswan Dam in Egypt, by preventing the flow of nutrients into the Mediterranean, virtually wiped out the native sardine industry, causing a loss to the country of some million dollars a year. It has also accelerated marine erosion of coastal Egypt, which now threatens the fertile delta lands.

Again in Peru, the offshore anchovy fisheries have been so grossly mismanaged by foreign companies that stocks have fallen drastically; while poor Peruvians suffer from lack of protein, most of the fish goes to Europe or North America where it is fed to livestock and pets.

The Kariba Dam in Central Africa was built without an adequate environmental survey of the lake basin, or of the places to which the local people were moved; fish yields dropped, waterweeds spread, and the tsetse flies flourished along the shores of the artificial lake.

The Aswan Dam led to the proliferation of a water snail that harbours a parasite carrying schistosomiasis, a particularly unpleasant and debilitating disease which has spread enormously since the dam was built.

Since the conference, Milton and a young Iranian from Commoner's team, Taghi Farvar, have been working together to produce a book based on the conference and a number of other case historial of ecological misfortune, nearly 200 in all. It has taken them three years to complete, and Milton expects it to have 'quite an impact'. He reckons that the evidence proves, in down-to-earth economic terms, that without ecological forethought development projects can do more harm than good.

It was this experience that led Milton into a critical study of the biggest, most ambitious and expensive development project of all time, the Mekong Project. Milton's outspoken attack on this has got him into hot water, I learned afterwards with many old-fashioned development people who have consistently played down the potential environmental impact of large dams on the Mekong.

'One of the most interesting aspects of our Conference to me personally was the accumulation of data about major river basin developments', said Milton. 'They seemed to combine a whole range of problems; engineering, pollution, fisheries, health, agriculture, social impact. The Smithsonian Institution in Washington got together a team to go out to the Mekong area in South East Asia for five months and investigate. In spent over half a year there organizing the research. What was exciting was that here, for once, was a chance for ecologists to be involved in a pre-investment survey. And its on such a mammoth scale; a multi-billion dollar project, involving a projected series of at least thirty, and probably over forty dams on the Mekong river and its tributaries, affecting several thousand square miles of South East Asia and the lives of over twenty million people.'

The Mekong river project was first dreamed up by the World Bank under a former director, Eugene Black, and presented to the public in 1964 by President Johnson as a positive alternative to the Vietnam war, from which both Hanoi and Saigon would benefit. It was hailed at the time as an indication of the generous and constructive instincts of the West towards the East. It is a co-operative venture, run from Bangkok by a committee consisting of the four nations directly concerned – South Vietnam, Cambodia, Laos, and Thailand – plus representatives of national and international agencies like AID, the Bank, the U.N., and other countries who decided to invest in it, such as Japan.

'The point about the Mekong river', Milton told me, 'is that life there is based largely on a traditional rice paddy culture, which is a beautifully worked-out system of land use depending on the natural flooding of the rivers. Basically what happens is that the rich sediment and silt left behind after the floods recede provide the basis for growing the rice, and also for excellent fisheries, which is where most of the local people get their protein supply. It's a very delicately balanced system, and it has proved itself over thousands of years. One thing that struck me very forcibly when I started looking into the planning was that no serious effort had been made to look at alternatives. The project went forward on the assumption that a series of large dams would be a good thing. Why hadn't the planners gone in first to study the region, the social system, the culture and values of the inhabitants? No one asked those twenty or thirty million people how they saw their needs and problems, or asked them what they wanted. How would they like to change? Do they want to change at all? Once we knew what they wanted, we should surely have looked at alternative ways of achieving development. We could have developed a series of fish-ponds and irrigation tanks, for instance, which are relatively cheap, easily manageable, give a high income, and provide much-needed protein as well as extra water for irrigation. We could have combined this with some investment in health measures, better sanitation, and some measure of population control. But by the time our group got there, the assumption had long been made that those dams were going to be built; the only questions were how many, where, and how.

'What shocked me was that no real effort had been made to consult the people affected. As with other big dam projects, the first that many of those people will probably know of it is when the flood waters begin to rise around their villages. Almost all the pre-investment feasibility work on the Mekong plan has been focused on economics and engineering. Our team was only there five months, but we began to have some idea of the kind of problems that are likely to arise. For instance, when you construct dams, you are bound to create a barrier to the migration of fish. When flooding is restricted, the lateral spread of the rivers is limited and the spawning of the fish is affected. Nobody has done any proper biological studies of the impact there'll be on the various fish species in the region.

'Also, the Great Lake of Cambodia is bound to be affected. This is one of the most incredible freshwater fisheries in the world. It is drained by a small river that connects with the Mekong, but at flood time a great pulse of water comes down the Mekong and reverses the flow of the smaller tributary, and the Great Lake expands to three or four times its size. Then spawning takes place, and the fish move into an area of flooded forest around the lake's edge, where they are protected from predators. Then, as the flood water recedes, the fish move back into the main body of the lake and also into the Mekong itself.

'There are thousands of people who live in permanent floating villages on this lake, in houseboats and on rafts, and they help to feed thousands more Cambodians. Dams on the major stream of the Mekong are bound to interfere with this great natural, cyclical rhythm of flooding and recession. Undoubtedly this is going to have a most drastic effect on the productivity of the fisheries and on the life of the whole area.'

Although most of Milton's working life is spent worrying about ecological problems in the Third World, he is preoccupied, like many other young Americans, with the subtle and nagging question of how to reflect concern over the environment in his own day to day life. He admits it is easier to feel in touch with nature if you are lucky enough, as he was, to grow up at least partly in the country. However, with the best will in the world, it is difficult, he feels, for urban or suburban kids to follow ecologically desirable practices, even if they do decide to opt for a commune in the wilds.

Milton spent a large part of his childhood on a remote eighteenth-century farm in New Hampshire, owned by his grandfather, where there were virtually no machines, and the family grew their own fruit and vegetables. 'We lived very close to the land', he said wistfully. 'I sometimes feel that if you don't grow up in that context, you don't ever get your head together.' Milton's family still owns the farm, and he goes back there whenever he can. He also tries to take off completely alone into the mountains somewhere for a few weeks each year.

He has mixed feelings about the efforts of young people to set up communes and live off the land; too many of them, in his view, are completely unrealistic. 'It's quite tragic. They want to get back to the land but they don't know what they're doing, and they make so many mistakes. I'd expect the great majority of these experiments to fail.' At the same time, he respects them for trying, and is conscious of the conflict in his own life. He is working for what he believes in, but lives in a way of which he disapproves. 'I drive a car. I fly around in aeroplanes. I use up power in my office and at home. I'm plugged into a whole system which I feel is wrong. I should really try to live in the country somewhere near Washington, with my own food sources and energy supply, using solar power and maybe a small hydropower system or a windmill.' The trouble is that, at the moment, it is the people who are opting out of society who are choosing to live in this way, not people like Milton who are trying to change it.

But Milton feels strongly that it is especially important for people like him to set an example. 'I'm still living within the system, I know, but in the last two or three years I've made quite a shift', he said earnestly. 'I've taken all my investments out of stocks and savings accounts, for instance, where my money can be used by forces and agencies of which I disapprove. Instead, I've put it into supporting environments which I feel are ecologically whole, or where I can get good land use methods established. About six years ago I joined with a group of friends and bought an island off the coast of Georgia. We've zoned it carefully, and anyone who builds a cottage there is asked to keep within very restricted guidelines. We apply ecological critieria to building materials, and, of course, we avoid all pollution. We keep all the beaches wild and untouched. If we hadn't bought it, the island would have got into the hands of a developer who wanted to put up hundreds of shoddy second-hand homes for rich people.'

Engaging as this project sounded, I could not help making the obvious point that not many people had the money to take out of investments in the first place, let alone buy an island and develop it in an ecological manner. Milton bristled. 'It really doesn't take that much money, when you think what people in this country spend on second or third cars or other appliances. In Maine it's possible to find a farm of several hundred acres for ten or fifteen thousand dollars. If you have fifteen or twenty people involved, that doesn't come very expensive. If you can buy a new Ford, you can also put that money into building an ecologically-balanced community.'

He has been thinking about producing a set of ecological guidelines for people who would like to know more about this kind of experiment, as a contribution towards making the growing commune movement more constructive. I pictured for a moment another conference, to be attended this time not by officials and professional ecologists but by young people from all over the world who were interested in finding an alternative life style based on ecological principles. If anyone could bring off such a project, I thought, it would be Milton who, for all his impressive qualifications and wide experience, seemed to me to have retained much of the directness and idealism of the younger generation. Like many of them, he has an intense interest in Eastern philosophy, which he maintains is fundamentally ecological.

'Zen Buddhism is very much opposed to making any distinction between man and nature', he said. 'Ecology and certain trends in oriental thought link up quite closely. I've been thinking of doing a book on it. I'd like to go to Japan and live there for a while and really work it out.'

I mentioned the irony that Japan seems to be the country where rapid industrial expansion has produced the most acute environmental problems of anywhere in the world.

'But that is perfectly explicable in terms of their political and economic history', said Milton quickly. 'It only happened when they opened up their society to Western European influences in the nineteenth century. However, the Zen tradition came from China many hundreds of years earlier.'

I asked him to explain more closely how Zen Buddhism and ecology converge.

'Zen teaches that we should see ourselves as part of the constant flow of matter and energy', he said. 'It says that there really is no distinction between the organism and its environment. Man is part of nature but western man has allowed his capacity for analytic thought to become too dominant. There is a basic paradox in the nature of thought itself. If I let myself think I am looking at a tree, it means I have decided that the tree and myself are separate entities. That's the first step towards manipulation of nature. The Zen teacher would say: see! The ecologist would point out how both man and tree are completely linked in an interdependent ecosystem.'

But surely, ecologists especially cannot afford to renounce man's capacity for applying his intelligence to the way nature works?

'Renounce our reason? No, of course not', said Milton. 'But we as organisms, and the environment, of which we are part, are in constant flux. Nothing remains constant, including ourselves. Our own concept of self relates more to the flow of past experience than to what we really are now. Similarly, all environments are dynamic systems, changing constantly through evolution and the introduction of new elements to each system. By the time we've finished studying a place, both it, and ourselves, will have changed. It is important for us to remember that rational thought is an imperfect tool – particularly when it leads us to manipulate environments and others as objects apart from us.'

POSTSCRIPT

WHILE I was working on this book, it became apparent that ecology, whatever else it was, was not a passing fad. Any subject that becomes instantly fashionable can be as instantly eclipsed; but ecology and the environmental movement are here to stay.

Two years after I discovered ecology, the environmentalists in America came close to getting a major underground nuclear explosion called off. In Britain, on a smaller scale, but no less significantly, the county of Cheshire appointed a full-time ecologist to the county council planning staff, and my local branch of Friends of the Earth, the American-inspired ecology action group now spreading through Europe, started to collect waste paper, free, for recycling.

These are hopeful signs. They indicate the ecological message is getting through where it counts, to ordinary people who, when they become critical, have the power to change things, which a group of scientists, however dedicated and influential, does not.

On one level, my investigations into the nature of ecology confirmed my initial feeling that it can be simply a new understanding of the inter-relatedness of everything. Like a religious revelation, once you have seen this, you cannot imagine how you were so blind before. If seventeen trees have to be cut down to make one ton of paper, then, of course, paper should be recycled. If birds and fish die from exposure to pesticides, then we should control their application. Once reminded that man is part of a complex, delicate natural system – a great web in which damage to one part is almost bound to shake the rest – you don't easily forget it. All the ecologists I talked to had this awareness, though some expressed it and applied it more readily than others.

All of them, too, were deeply fascinated by the intricate mechanisms of nature. It turned out to be much easier than I had expected to appreciate the detail of most ecological work. The substance of ecology, I learned, is endlessly fascinating, and not at all a closed world for specialists alone. If you enjoy walking along a beach inspecting rock pools, or recognizing wild flowers, or gardening, or keeping fish, ecology is probably already part of your thinking. For this reason, it seems to me that the enthusiasm of amateur naturalists could be used, much more effectively than it is at the moment, to increase the stock of ecological knowledge, and to spread awareness of how vulnerable nature is. Children can be very good at it, as a recent project in England proved, when thousands of children spent their summer monitoring water pollution in their local streams and ponds.

But to me, the real revelation, and one that is still, I suspect, not widely enough appreciated, was the potential of ecology for changing our whole system. There seem to be certain imperatives built into the ecological approach which, if they are carried through, will have the most drastic effect on our habits. For example, waste must stop, so recycling must become official. We must again manufacture goods that are made to last. New industries and technologies will need wholly new criteria. If supersonic aircraft might set off dangerous chemical reactions in the upper atmosphere, they must not fly, and ought never to have been built. If chemical leaks and dumping can kill lakes, foul rivers, and poison oceans, they must be stopped. Someone will have to pay.

All these imperatives involve decisions which will mean changes, sometimes uncomfortable ones, for us all. Plastic bags are useful in the kitchen, but housewives should do without them. Cars are a great convenience, but it looks as if we must severely restrict their use. If the population increase cannot be slowed down fast enough by social and economic improvements, then perhaps, we must all be limited to two children.

To pursue this line of thought is to realize that ultimately ecology, if it is to work, must involve a radical change in our economic and social priorities. It means that ecological sanity must replace profit and productivity as our goal.

The ecologists can provide the information on which the decisions to change can be based but they are not going to take the decisions for us. It is true that the ecologists should be listened to and consulted more than they are at present; equally, it seems to me, too few ecologists are prepared to make a public fuss about the dangers they foresee. If the world waits until many of these threats materialize, it will be too late. Already there are indications that water pollution can cause new forms of disease; it has been demonstrated, for instance, that people who swim in the Mediterranean off France, Spain, and Italy are twice as likely to contract certain infections as people who do not. What are the European ecologists doing about that?

Inevitably, the role of the ecologist in sorting out the environmental mess is bound to vary according to individual temperament. We certainly need more ecologists; we certainly need more activist ecologists; but we also need more economists, politicians, industrialists, and ordinary citizens who think ecologically.

Ecology is not, however, a subject which provides easy answers, instant remedies, or short cuts. The work is always minute and long-term and the results are often confusing. The great D.D.T. controversy is a good example of this. There has been an enormous amount of discussion and research on the safety of D.D.T. Yet it is extraordinarily difficult to get a clear answer to the question: should D.D.T. now be banned everywhere? There is violent disagreement among scientists and development agency administrators as to whether or not D.D.T. is actually harmful to people, and whether or not it impedes photosynthesis, the most vital of all natural processes, on which the world depends for oxygen. The best one can do is to state that there seems to be at least a tacit acknowledgement by everyone that it would probably be as well if D.D.T. was phased out and replaced, as soon as possible, by other methods. The D.D.T. problem is so intransigent largely because so many countries rely on it for saving lives and improving public health, at any rate in the short term. D.D.T. had already become part of the accepted range of chemicals added to the environment before anyone realized its disadvantages.

The single most crucial precept I learned from the ecologists was this: we must ensure that the products of modern science and technology, whether dams, or plastics, or chemicals, are checked out in advance against ecological criteria. It is useful to remember here that environmentalists are beginning to object to the term 'side-effects'. Air pollution is as much an effect of the internal combustion engine as is getting from A to B. Junk on the world's beaches is an effect of the invention of plastics, not a mere 'side-effect'.

On the recurrent question of how urgent the environmental crisis really is, of whether we can set a time limit on the survival of civilization as we know it, not even the ecologists can give a precise answer. What is clear is that if you think of the planet as a spaceship, a lot of warning lights are flashing. The general impression I got was that unless we change our ways before the end of this century, we are going to be in serious trouble.

As for the charge so often levelled at the more impassioned ecologists, that they are overdramatizing the immediate dangers of the world environmental situation, in my experience the more active, crusading ecologists are driven to do so by the general apathy and inertia still surrounding the subject. I found all the ecologists reasonably optimistic that we can still get things straight – *if* we don't waste any more time.